Microbial Adhesion and Invasion

Magnus Hook Lech Switalski
Editors

Microbial Adhesion and Invasion

With 36 Illustrations

Springer-Verlag
New York Berlin Heidelberg London Paris
Tokyo Hong Kong Barcelona Budapest

Magnus Hook, Ph.D.
Department of Biochemistry
University of Alabama
Birmingham, AL 35294
USA

Lech Switalski, Ph.D.
Departments of Microbiology and
Biochemistry, and Periodontics
School of Dental Medicine
University of Pittsburgh
Pittsburgh, PA 15261
USA

Library of Congress Cataloging-in-Publication Data
Microbial adhesion and invasion : proceedings of the fifth symposium
initiated and sponsored by the Department of Biochemistry, Schools
of Medicine and Dentistry, University of Alabama at Birmingham, held
at Edgewater Beach Resort, Panama City, Florida, October 25–27, 1990
Lech Switalski and Magnus Hook, editors.
p. cm.
Includes bibliographical references and index.
ISBN 0-387-97815-1. — ISBN 3-540-97815-1.
1. Microorganisms—Adhesion—Congresses. 2. Microbial
invasiveness—Congresses. I. Hook, Magnus. II. Switalski, Lech.
III. University of Alabama at Birmingham. Dept. of Biochemistry.
[DNLM 1. Bacteria—pathogenicity—congresses. 2. Cell Adhesion
—congresses. QW 730 M6255 1990]
QR96.8.M52 1992
606′.01—dc20
DNLM/DLC
for Library of Congress 92-2186

Printed on acid-free paper.

Production managed by Hal Henglein; manufacturing supervised by Genieve Shaw.
Camera-ready copy prepared by the editors.
Printed and bound by Edwards Brothers, Inc., Ann Arbor, MI.
Printed in the United States of America.

9 8 7 6 5 4 3 2 1

ISBN 0-387-97815-1 Springer-Verlag New York Berlin Heidelberg
ISBN 3-540-97815-1 Springer-Verlag Berlin Heidelberg New York

Preface

This book contains the proceedings of the symposium on Microbial Adhesion and Invasion.

The meeting was the fifth in a series of symposia initiated and sponsored by the Department of Biochemistry, University of Alabama at Birmingham. We gratefully acknowledge the generous financial support of the following institutions and companies.

- Department of Biochemistry, University of Alabama at Birmingham, Birmingham, Alabama, USA

- National Institutes of Health, Bethesda, Maryland, USA

- Alfa Laval International, AB, Tumba, Sweden

- Procter and Gamble, Cincinnati, Ohio, USA

- Amgen Inc., Thousand Oaks, California, USA

- Johnson and Johnson, New Brunswick, New Jersey, USA

We would like to express our appreciation to Agneta Hook, Timo Kostiainen, Mary Homonylo McGavin, Martin McGavin, Cindy Patti, Joe Patti and Anna-Marja Saamanen for their time and hard work making this symposium a success; and special thanks to Kay Cooper and Wade Butcher for help with the management of the symposium and editorial assistance with this book.

Magnus Hook
Department of Biochemistry
University of Alabama at Birmingham

Lech M. Switalski
Departments of Microbiology and Biochemistry, and Periodontics
University of Pittsburgh

Contents

Participants

Soman N. Abraham, University of Tennessee, 956 Court Avenue, Memphis, TN 38163, USA

Richard S. Ajioka, Research Institute of Scripps Clinic, 10666 North Torrey Pines Road, Mail Code MB-4, La Jolla, CA 92037, USA

Par Aleljung, University of Lund, Medical Microbiology, Solvegatan 23, S 223 62 Lund, Sweden

Bradley L. Allen, University of Iowa, 3-301 BSB, Iowa City, IA 52242, USA

Roger Allen, University of Alabama at Birmingham - Periodontics, UAB Station-SDB 710, Birmingham, AL 35294, USA

Heather Allison, University of Florida, Box J-266, JHMHC, Gainesville, FL 32610, USA

Earl J. Bergey, SUNY at Buffalo, Foster Hall 109, Oral Biology, Buffalo, NY 14214, USA

Debra E. Bessen, Rockefeller University, 1230 York Avenue, Box 276, New York, NY 10021, USA

Gary K. Best, Medical College of Georgia, Cellular & Molecular Biology, Augusta, GA 30912, USA

Norma Best, VA Medical Center, Augusta, GA 30912, USA

Kristin A. Birkness, Centers for Disease Control, Bldg 1, Rm 2221, Mailstop D11, Atlanta, GA 30333, USA

Robert D. Blalock, University of Tennessee, 858 Madison Avenue, Room 601, Memphis, TN 38163, USA

Lawrence B. Blyn, University of Utah, School of Medicine, Salt Lake City, UT 84132, USA

Michael Brennan, CBER, FDA, Building 29, Room 418, 8800 Rockville Pike, Bethesda, MD 20892, USA

Robert E. Briggs, National Animal Disease Center, Box 70, Dayton Road, Ames, IA 50036, USA

David E. Briles, University of Alabama at Birmingham - Microbiology, UAB Station - SDB 801, Birmingham, AL 35294, USA

Michael S. Bronze, VA Medical Center (111M), 1030 Jefferson Avenue, Memphis, TN 38104, USA

Robert C. Brown, University of Tennessee, Microbiology & Immunology, 858 Madison Avenue, Memphis, TN 38163, USA

Clayton A. Buck, The Wistar Institute, 3601 Spruce Street, Philadelphia, PA 19104, USA

Wade Butcher, University of Pittsburgh, Microbiology and Biochemistry, and Periodontics, 566 Salk Hall, Pittsburgh, PA 15261, USA

Thomas B. Buxton, VA Medical Center, Infectious Diseases (111G), Augusta, GA 30910, USA

Jerry M. Buysse, Walter Reed Army Institute of Research, Washington, DC 20307, USA

J. Robert Cantey, VA Medical Center (151), 109 Bee Street, Charleston, SC 29403, USA

Frederick J. Cassels, Walter Reed Army Research Institute, Building 40, Room 2088, Washington, DC 20307, USA

Sukla B. Chattopadhyay, University of Alabama at Birmingham - Microbiology, UAB Station- BHSB 957, Birmingham, AL 35294, USA

Joseph C.R. Chen, University of Rochester, 601 Elmwood Avenue, Box 672, Rochester, NY 14642, USA

Gursharan S. Chhatwal, Arbeits Gruppe Der GRF at Technical University, Konstantin-Uhde Strasse 5, 3300 Braunschweig, Germany

Pawel Ciborowski, University of Lund, Medical Microbiology, Solvegatan 23, Lund, S 223 62, Sweden

Richard J. Colonno, Virus and Cell Biology, Merck Sharp & Dohme Research Laboratories, West Point, PA 19486, USA

Harry S. Courtney, VA Medical Center (151M), 1030 Jefferson Avenue, Memphis, TN 38104, USA

Madeleine W. Cunningham, Microbiology and Immunology, University of Oklahoma, Health Sciences Center, P.O. Box 26901, Oklahoma City, OK 73190, USA

James B. Dale, VA Medical Center (11A), 1030 Jefferson Avenue, Memphis, TN 38104, USA

Donald R. Demuth, University of Pennsylvania, Levy Bldg - Room 383, Philadelphia, PA 19104, USA

Michael S. Donnenberg, University of Maryland School of Medicine, 10 South Pine Street, Baltimore, MD 21201, USA

Wayne Duck, University of Alabama at Birmingham - Periodontal, UAB Station-SDB 710, Birmingham, AL 35294, USA

James P. Dudley, UCLA, Room 62, 158 CHS, Los Angeles, CA 90024, USA

Eric A. Elsinghorst, Walter Reed Army Institute of Research, Bacterial Immunology, Washington, DC 20307, USA

Monica M. Farley, Research Service (151), VA Medical Center, 1670 Clairmont Road, Decatur, GA 30033, USA

Barry S. Fields, Centers for Disease Control, Mail Stop CO2, Atlanta, GA 30333, USA

Vincent A. Fischetti, The Rockefeller University, 1230 York Avenue, New York, NY 10021, USA

Jonathan Fletcher, University of Liverpool, P.O. Box 147, Liverpool, L69 3BX, United Kingdom

Glynn H. Frank, National Animal Disease Center, P.O. Box 70, Ames, IA 50010, USA

Craig Franklin, W247 Veterinary Medicine, University of Missouri, Columbia, MO 65211, USA

Nancy E. Freitag, Microbiology, School of Medicine, University of Pennsylvania, Philadelphia, PA 19104-6076, USA

Gunnar Froman, University of Uppsala, Clinical Bacteriology, Uppsala S-751 22, Sweden

Janina Goldhar, Sackler School of Medicine, Tel Aviv University, Tel Aviv 69978, Israel

Eric A. Goulbourne, Procter & Gamble, Miami Valley Lab, P.O. Box 398707, Cincinnati, OH 45239, USA

Anthony G. Gristina, Musculoskeletal Sciences, Research Institute, 2190 Fox Mill Road, Herndon, VA 22071, USA

Sivashankarappa Gurusiddappa, University of Alabama at Birmingham - Biochemistry, UAB Station-BHSB 509, Birmingham, AL 35294, USA

Josee Harel, Montreal University, Veterinary School, CP 5000, St. Hyacinthe, Quebec, Canada 52S 7C6

J. Olu Hassen, Washington University, Campus Box 1137, St. Louis, MO 63130, USA

David L. Hasty, VA Medical Center, Research Service (151), 1030 Jefferson Avenue, Memphis, TN 38163, USA

Victor B. Hatcher, Albert Einstein College of Medicine, 101 East 210th Street, Bronx, NY 10467, USA

Linda Hazlett, Wayne State University, School of Medicine, Anatomy & Cell Biology, Detroit, MI 48201, USA

Susan T. Hingley, Hahnenmann University, Microbiology & Immunology, MS 410, Broad and Vine Streets, Philadelphia, PA 19102, USA

Susan Hollingshead, University of Alabama at Birmingham - Microbiology, UAB Station - SDB 801, Birmingham, AL 35294, USA

Kathryn V. Holmes, Pathology, Uniformed Services University of the Health Sciences, 4301 Jones Bridge Road, Bethesda, MD 20889-1640, USA

Anders Holmgren, Swedish University of Agricultural Sciences, Uppsala Biomedical Center, P.O. Box 590, S 751 24 Uppsala, Sweden

Magnus Hook, University of Alabama at Birmingham - Biochemistry, UAB Station-BHSB 508, Birmingham, AL 35294, USA

Agneta Hook, University of Alabama at Birmingham - Biochemistry, UAB Station-BHSB 593, Birmingham, AL 35294, USA

Sophia Kathariou, University of Hawaii, Microbiology, 2538 The Mall, Snyder Hall 207, Honolulu, HI 96822, USA

Melissa R. Kaufman, University of Tennessee, 858 Madison Avenue, Microbiology & Immunology, Rm 601, Memphis, TN 38163, USA

C. Harold King, Centers for Disease Control, Respiratory Diseases Branch, Mail Stop C0-2, Atlanta, GA 30333, USA

Per Klemm, Technical University of Denmark, Building 222, DK-2800, Lyngby, Denmark

Francis W. Klotz, Walter Reed Institute of Research, Building 40, Room 1015, Washington, DC 20307, USA

Timo K. Korhonen, University of Helsinki, Mannerheimintie 172, Helsinki SF 00300, Finland

Timo Kostiainen, University of Alabama at Birmingham - Biochemistry, UAB Station-BHSB 505, Birmingham, AL 35294, USA

Malak Kotb, VA Medical Center (151), 1030 Jefferson Avenue, Memphis, TN 38104, USA

Danuta Krajewska-Pietrasik, University of Alabama at Birmingham - Biochemistry, UAB Station-BHSB 505, Birmingham, AL 35294, USA

Howard C. Krivan, 300 Professional Drive, Suite 100, Gaithersburg, MD 20879, USA

Meta Kuehn, Washington University, Molecular Biology, Box 8230, 660 South Euclid Avenue, St. Louis, MO 63110, USA

Pentti Kuusela, University of Helsinki, Haartmaninkatu 3, Helsinki, SF 00290, Finland

Marilyn Lantz, University of Pittsburgh, Periodontics, 3501 Terrace Street, Pittsburgh, PA 15261, USA

Elizabeth Leininger, Food and Drug Administration, 8800 Rockville Pike, HFB 670, Bethesda, MD 20892, USA

Robert D. Leunk, Procter & Gamble, Miami Valley Labs, P.O. Box 398707, Cincinnati, OH 45239, USA

Henrik Linder, University of Goteborg, Guldhedsgatan 10A, Goteborg S 413 46, Sweden

Richard Lottenberg, University of Florida, JHMHC-Hematology, Box J-227, Gainesville, FL 32610, USA

Franklin D. Lowy, Montefiore Hospital, 111 East 210th Street, Bronx, NY 10467, USA

Carol W. Maddox, University of Missouri, 1600 East Rollins Street, Columbia, MO 65211, USA

Berenice Madison, University of Tennessee, 858 Madison Avenue, Memphis, TN 38163, USA

Sourindra N. Maiti, University of Montreal, CP 5000 St. Hyacinthe, Quebec J25 7C6, Canada

Sou-ichi Makino, Max Planck Institute, Spemannstrasse 34, Tubingen, D 7400, Germany

Daniel Malamud, University of Pennsylvania, School of Dental Medicine, 4001 Spruce Street, Philadelphia, PA 19104, USA

G. Lynn Marks, University of South Alabama, LMB Building, Mobile, AL 36688, USA

Carl F. Marrs, University of Michigan, 109 Observatory Street, Ann Arbor, MI 48109, USA

Robert Masure, Rockefeller University, 1230 York Avenue, New York, NY 10021, USA

Kalai Mathee, University of Tennessee, 858 Madison Avenue, Room 711, Memphis, TN 38163, USA

Martin McGavin, University of Alabama at Birmingham - Biochemistry, UAB Station-BHSB 510, Birmingham, AL 35294, USA

Mary McGavin, University of Alabama at Birmingham - Biochemistry, UAB Station-BHSB 509, Birmingham, AL 35294, USA

John J. Mekalanos, Microbiology and Microbial Genetics, Harvard Medical School, 25 Shattuck Street, Boston, MA 02115, USA

Joseph M. Merrick, Microbiology, SUNY at Buffalo, Buffalo, NY 14214, USA

Jeffery Miller, Microbiology and Immunology, UCLA School of Medicine, Center for Health Sciences, 10833 Le Conte Avenue, Los Angeles, CA 90024-1747, USA

Clifford S. Mintz, University of Miami, School of Medicine, P.O. Box 016960 (R138), Miami, FL 33101, USA

Issei Nakayama, Nihon University, 1-8-13 Kanda Surugadai, Chiyoda-ku, Tokyo 101, Japan

Xavier Nassif, Research Institute of Scripps, Molecular Biology MB4, 10666 N. Torrey Road, La Jolla, CA 92037, USA

Nancy Ness Nichols, University of Iowa, Microbiology, 3-301 Bowen Science Bldg., Iowa City, IA 52242, USA

Wade A. Nichols, University of Iowa, Microbiology, 3-301 Bowen Science Bldg., Iowa City, IA 52242, USA

Staffan J. Normark, Washington University, School of Medicine, 660 South Euclid Avenue, Box 8230, St. Louis, MO 63116, USA

Bogdan Nowicki, University of Texas, Medical Branch of Galveston, Obstetrics & Gynecology, Galveston, TX 77550, USA

Orjan Olsvik, Norwegian College, Veterinary Medicine, P.O. Box 8146 DEP, 0033 Oslo 1, Norway

Linda M. Parsons, New York State Department of Health, Wadsworth Center for Laboratories & Research, P.O. Box 509, Albany, NY 12201, USA

Joseph Patti, University of Alabama at Birmingham - Biochemistry, UAB Station-BHSB 593, Birmingham, AL 35294, USA

Joel Peek, University of Tennessee, Microbiology and Immunology, 858 Madison Avenue, Memphis, TN 38163, USA

Kaety Plos, Gothenburg University, Guldhedsgatan 10A, Goteborg, 41346, Sweden

Daniel A. Portnoy, Microbiology, 209 Johnson Pavilion, University of Pennsylvania, Philadelphia, PA 19104-6076, USA

Frederick D. Quinn, Centers for Disease Control, Molecular Biology Laboratory, 1-2225 (D11), Atlanta, GA 30333, USA

Vincent Racaniello, Microbiology, Columbia University College of Physicians and Surgeons, 701 West 168th Street, New York, NY 10032, USA

Katherine Rostand, University of Alabama at Birmingham - Biochemistry, UAB Station- BHSB 565, Birmingham, AL 35294, USA

Richard F. Rest, University of Michigan, Microbiology, 6643 Medical Sciences #2, Ann Arbor, MI 48109, USA

James A. Roberts, Tulane University, Delta Primate Center, 18703 3 Rivers Road, Covington, LA 70433, USA

Frank G. Rodgers, University of New Hampshire, Spaulding Life Science Center, Durham, NH 03824-3544, USA

Michael W. Russell, University of Alabama at Birmingham - Microbiology, UAB Station-ZRB 418, Birmingham, AL 35294, USA

Cecilia Ryden, Uppsala University, Medical & Physiological Chemistry, Box 575 Biomedical Center, Uppsala S 751 23, Sweden

Anna-Marja Saamanen, University of Alabama at Birmingham - Biochemistry, UAB Station-BHSB 593, Birmingham, AL 35294, USA

Philippe J. Sansonetti, Unite de Pathogenie, Microbienne Moleculaire, Institut Pasteur, 28, Rue du Dr. Roux, 75724 Paris Cedex 15, France

Sergio Schenkman, New York University, Medical Center, MS 131 - Pathology, 550 1st Avenue, New York, NY 10016, USA

Barbara Ann Sanford, University of Texas, Health Science Center, Microbiology, 7703 Floyd Curl Drive, San Antonio, TX 78284, USA

David Senior, University of Florida, Box J-126 HSC, Gainesville, FL 32607, USA

Katherine Shih, Sola/Barnes-Hind, Inc., 810 Kifer Road, Sunnyvale, CA 94068, USA

Warren Simmons, University of Alabama at Birmingham - Microbiology, 501 Volker Hall, Birmingham, AL 35294, USA

Lynn Slonim, Washington University, School of Medicine, Molecular Biology, Box 8230, 660 South Euclid Avenue, St. Louis, MO 63110, USA

Samuel D. Smith, Children's Hospital of Pittsburgh, Pediatric Surgery, 3705 Fifth Avenue at DeSoto Street, Pittsburgh, PA 15213-2583, USA

Patricia G. Spear, Microbiology and Immunology, Northwestern University Medical School, 303 East Chicago Avenue, Chicago, IL 60611, USA

Pietro Speziale, Biochemistry, University of Pavia, Via Bassi, 21 27100 Pavia, Italy

Christian Stadtlaender, University of Alabama at Birmingham - Microbiology, 533 Volker Hall, UAB Station, Birmingham, AL 35294, USA

David S. Stephens, Emory University, School of Medicine, 69 Butler Street, Atlanta, GA 30303, USA

Bruce A.D. Stocker, Microbiology and Immunology, Stanford University School of Medicine, Stanford, CA 94305-5402, USA

Soila Sukupolvi, National Public Health Institute, Mannerheimintie 166, Helsinki, SF 00300, Finland

Paul M. Sullam, VA Medical Center 113A, 4150 Clement Street, San Francisco, CA 94121, USA

Lech M. Switalski, University of Pittsburgh, Microbiology and Biochemistry, and Periodontics, 566 Salk Hall, Pittsburgh, PA 15261, USA

Edward Swords, University of Alabama at Birmingham - Microbiology, UAB Station-SDB 801, Birmingham, AL 35294, USA

Ann Mari Tarkkanen, University of Helsinki, Mannerheimintie 172, Helsinki SF 00300, Finland

Rebecca C. Tart, Bowman Gray School of Medicine, 300 South Hawthorne Road, Winston-Salem, NC 28677, USA

David N. Taylor, Center for Vaccine Development, 10 South Pine Street, Baltimore, MD 21201, USA

Ronald K. Taylor, University of Tennessee, 858 Madison Avenue, Memphis, TN 38163, USA

Robert W. Tolan, Washington University, School of Medicine, Molecular Biology & Pediatrics, St. Louis, MO 63110, USA

Anna L. Trifillis, University of Maryland, MSTF 7-16, 10 South Pine Street, Baltimore, MD 21201, USA

Elaine Tuomanen, Rockefeller University, 1230 York Avenue, New York, NY 10021, USA

Loek van Alphen, University of Amsterdam, Medical Microbiology, Room L1-162, Meiberqdeef 15, Amsterdam NL-1105AZ, The Netherlands

Ivo van de Rijn, Wake Forest University, 300 South Hawthorne Road, Winston-Salem, NC 27103, USA

Marieke van Ham, Max Planck Institute, Spemannstrasse 34, Tubingen, 7400 Germany

Ritva Virkola, University of Helsinki, Mannerheimintie 172, Helsinki SF 00300, Finland

Torkel Wadstrom, University of Lund, Medical Microbiology, Solvegatan 23, S 223 62 Lund, Sweden

Yngvild Wasteson, Microbiology and Immunology, Norwegian College of Veterinary Medicine, P.O. Box 8146, DEP 0033 Oslo 1, Norway

Haruo Watanabe, National Institute of Health, Bacteriology, 2-10-35 Kamiosaki, Shinagawa-ku, Tokyo 141, Japan

Carol L. Wells, University of Minnesota, Laboratory Medicine and Pathology, Box 198, Mayo Building, Minneapolis, MN 55455, USA

Benita Westerlund, University of Helsinki, Mannerheimintie 172, Helsinki, SF 00300, Finland

Elizabeth "Libby" White, Centers for Disease Control, Building 1, Room 2316 D-17, 1600 Clifton Road, Atlanta, GA 30333, USA

Christine White-Ziegler, University of Utah, Medical School, Salt Lake City, UT 84132, USA

Mary Jo Wick, Washington University, Medical Center, 660 South Euclid Avenue, Box 8230, St. Louis, MO 63116, USA

Phletus P. Williams, National Animal Disease Center, P.O. Box 79, Ames, IA 50010, USA

David O. Wood, University of South Alabama, College of Medicine, Laboratory of Molecular Biology, Mobile, AL 36688, USA

Emiko Yamaji, Nihon University, 1-8-13 Kanda Surugadao, Chiyoda-ku Tokyo 101, Japan

Janet Yother, University of Alabama at Birmingham - Microbiology, UAB Station-SDB 802, Birmingham, AL 35294, USA

Dick Zoutman, University of Alberta, M330 Biological Science Building, Edmonton, Alberta, T6G 2E9, Canada

Edwin H. Beachey
In Memoriam

Dr. Edwin H. Beachey, who made major scientific contributions to the field of Infectious Diseases and Immunology, died October 27, 1989 in Memphis, Tennessee. He was 55 years old. At the time of his death, Dr. Beachey was Professor of Medicine and Microbiology, and Chief of the Division of Infectious Diseases, at the University of Tennessee College of Medicine. He was also Associate Chief of Staff for Research and Development and Director of the Infectious Disease Research Program at the Memphis Veterans Administration Medical Center.

Ed was best known for his pioneering work in the field of bacterial adherence, with milestone observations involving both Gram-positive and Gram-negative bacteria. These observations contributed to the understanding of bacterial pathogenesis, including the molecular basis for the adherence of bacteria to host cells, and the molecular mechanisms of streptococcal virulence. He was also known for his fundamental contributions to the immunology of group A streptococci, particularly the immunopathogenesis of rheumatic fever and the development of novel synthetic and recombinant vaccines.

Ed was one of the organizers of the previous Gulf Shores Symposium in 1988. Although we are deeply saddened by the loss of a great scientist and a dear friend, we are grateful for the legacy he left us which continues to contribute to our understanding of microbial adhesion and invasion. These proceedings contain his last review on the development of vaccines against group A streptococci.

Cell Surface Carbohydrates as Adhesion Receptors for Many Pathogenic and Opportunistic Microorganisms

Howard C. Krivan, Laura Plosila, Lijuan Zhang, Valerie Holt, and Mamoru Kyogashima

Introduction

The human cell surface is covered with a layer of membrane-bound carbohydrate in the form of oligosaccharides linked covalently to lipids and proteins. The carbohydrate moieties of these glycolipids and glycoproteins form a network of outer surface projections that can serve as adhesion sites, i.e. receptors, for a variety of infectious agents which come into contact with the host cell. In many cases, the adhesion site involves a particular carbohydrate sequence which encodes a very precise binding domain that directly contributes to specificity and tissue tropism. Although carbohydrates have been recognized as possible receptors for infectious agents for quite some time (Beachey, 1981; Jones and Isaacson, 1985), it has only been within the past few years that several of these sequences have been identified and reported as possible adhesion receptors, and most of this work has been restricted to glycolipids (reviewed by Karlsson, 1989). Perhaps the major reason for this advancement is that glycolipids contain one carbohydrate moiety per molecule, allowing for less complicated analysis and identification of the carbohydrate consensus sequences involved in specific binding. Glycoproteins, on the other hand, contain several N- and/or O-linked oligosaccharides and can be clustered within heavily glycosylated regions of the peptide chain. Such microheterogeneity makes for a more difficult study in defining exactly what particular carbohydrate sequence contributes to microbial receptor specificity. In addition to the technical advantages of studying glycolipids as microbial receptors, the role they play as adhesion receptors in infection may be very important to a particular incoming pathogen. Although glycolipids and glycoproteins are embedded in the plasma membrane architecture formed by phospholipids, sphingomyelin and cholesterol, the carbohydrate moieties of the glycolipids are physically closer to the membrane surface than those of glycoproteins. Thus, bacteria bound to glycolipid receptors are intimately associated with the host cell surface during colonization or infection and such intimacy may provide a better nutritional environment for a pathogen. In this paper, we report that there are at least three major classes of glycolipids that are involved in the adhesion of microorganisms to cells: sulfatides, which are receptors for mycoplasmas, and the ganglio- and lacto-series glycolipids which are receptors for several genera of pathogenic and opportunistic bacteria. The identification of the consensus sequences involved in binding of microorganisms and the exploitation of these receptors in biotechnology is discussed.

Members of the Order *Mycoplasmatales* Bind Specifically to Sulfatides

Mycoplasmas are small, cell wall-less procaryotic microorganisms. They are ubiquitous in nature and have been isolated from many animal and plant species. Some species are pathogenic for humans while others in the *Mycoplasmatales* do not cause disease, but are notorious for infecting cell cultures (McGarrity *et al.*, 1988). We have previously demonstrated that a virulent strain of the human pathogen *M. pneumoniae* specifically bound to sulfatide and other sulfated glycolipids, such as seminolipid and lactosylsulfatide (for structures see Table 1), and that the consensus binding sequence was a terminal Gal($3SO_4$)β1-residue (Krivan *et al.*, 1989). Recently, this same binding specificity was reported for mycoplasmas involved in animal and human infertility (Lingwood *et al.*, 1990). We also showed that the *M. pneumoniae* receptor occurs in substantial amounts in human trachea, lung, and WiDr cells, and that specific binding could be inhibited by using the receptor analogue dextran sulfate (Krivan *et al.*, 1989). To further examine the possible role of sulfated glycolipids as adhesion receptors, other species of the mycoplasma group were studied for binding to glycolipids.

Incubation of ^{3}H-labeled *M. hominis*, an organism frequently isolated from the human genital tract, with various glycolipids resolved on thin layer chromatograms was used to determine the carbohydrate binding specificity of the organism. As shown by an autoradiogram (Fig. 1B) compared with an identical thin layer plate visualized with orcinol reagent (Fig. 1A), *M. hominis* bound avidly to authentic sulfatide, detecting about 0.06 mg of this glycolipid (Fig. 1B, lane 6). *M. hominis* also bound to other sulfated glycolipids including lactosyl sulfatide and seminolipid, which contain the same terminal Gal($3SO_4$)β1-residue as sulfatide (Table 1). Interestingly, *M. hominis* also bound weakly to galactosylceramide (Fig. 1B, lane 1), but not to other neutral or acidic glycolipids tested, including the gangliosides GM3, GM2, GM1, sialylparagloboside, GD_3, GD_{1a}, GD_{1b} and GT_{1b} (Fig. 1B, lane 1; Table 1).

Table 1. Glycolipids Tested for Ability to Bind Mycoplasmas on Thin-layer Chromatograms.

Glycolipid[a]	Structure	Binding[b]
Sulfatide	Gal(3SO$_4$)β1-1Cer	+++
Lactosylsulfatide	Gal(3SO$_4$)β1-4Glcβ1-1Cer	+++
Seminolipid	Gal(3SO$_4$)β1-3alkylacylglycerol	+++
Galactosylceramide (CMH)	Galβ1-1Cer	+
Lactosylceramide (CDH)	Galβ1-4Glcβ1-1Cer	-
Globotriaosylceramide (CTH)	Galα1-4Galβ1-4Glcβ1-1Cer	-
Globotetraosylceramide (GL4)	GalNAcβ1-3Galα1-4Galβ1-4Glcβ1-1Cer	-
Asialo-GM 2	GalNAcβ1-4Galβ1-4Glcβ1-1Cer	-
Asialo-GM1	Galβ1-3GalNAcβ1-4Galβ1-4Glcβ1-1Cer	-
GM3	NeuAcα2-3Galβ1-4Glcβ1-1Cer	-
GM2	GalNAcβ1-4[NeuAcα2-3]Galβ1-4Glcβ1-1Cer	-
GM1	Galβ1-3GalNAcβ1-4[NeuAcα2-3]Galβ1-4Glcβ1-1Cer	-
Sialylparagloboside	NeuAcαGalβ1-4GlcNAcβ1-3Galβ1-4Glcβ1-1Cer	-
GD3	NeuAcα2-8NeuAcα2-3Galβ1-4Glcβ1-1Cer	-
GD1a	NeuAcα2-3Galβ1-3GalNAcβ1-4[NeuAcα2-3Galβ1-4Glcβ1-1Cer	-
GD1b	Galβ1-3GalNAcβ1-4[NeuAcα2-8NeuAcα2-3]Galβ1-4Glcβ1-1Cer	-
GT1b	NeuAcα2-3Galβ1-3GalNAcβ1-4[NeuAcα2-8NeuAcα2-3Galβ1-4Glcβ1-1Cer	-

[a]Trivial names and structures are represented according to recommendations in Reference 14 and references cited therein; cer, ceramide; CMH, ceramide monohexoside; CDH, ceramide dihexoside (lactosylceramide); CTH, ceramide trihexoside (globotriaosylceramide); GL4, globotetraosylceramide.

[b]Negative binding (-) indicates no binding to 2 μg of glycolipid and positive binding to less than 0.04 μg (+++), and 0.5-1 μg (+).

Because dextran sulfate was found to be a potent inhibitor of binding of *M. pneumoniae* to WiDr cells and to purified sulfatide (Krivan *et al.*, 1989), this receptor analogue and other anionic polysaccharides were tested for inhibition of *M. hominis* binding to sulfatide immobilized in microtiter plates (Table 2). Dextran sulfate was also found to be the most potent inhibitor of *M. hominis* binding to 1 mg of sulfatide with 50% inhibition obtained at 5 ug/ml, whereas the sulfated fucan, fucoidin, was a weak inhibitor, and dextran and several other polysaccharides tested were inactive (Table 2). Similar results in binding specificity and inhibition were obtained with a variety of other mycoplasmas listed in Table 3, including *Acholeplasma laidlawii* and *Ureaplasma urealyticum*. Because these organisms are very common contaminants of serum, media and cell cultures, it was of interest to determine whether dextran sulfate immobilized on beads could trap mycoplasmas and thus remove them from contaminated biofluids, such as fetal calf serum or cell culture media. ^{3}H-labeled *M. orale* was passed through columns containing immobilized dextran sulfate or other ligands to

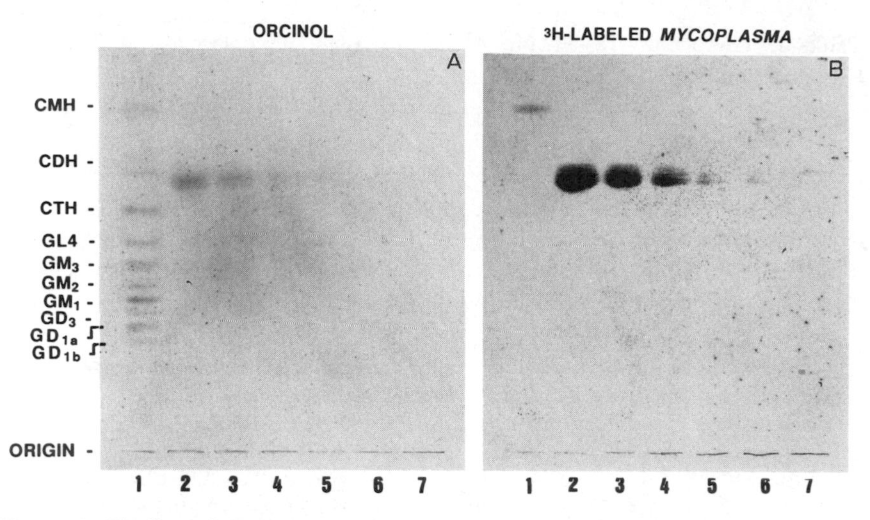

Figure 1: Binding of 3H-labeled *M. hominis* to standard glycolipids and authentic sulfatide separated by thin layer chromatography. Glycolipids were chromatographed on aluminum-backed silica gel high performance thin layer chromatography plates developed in chloroform, methanol, 0.25% KCl in water, 5:4:1. Plates were coated with 0.1% polyisobutylmethacrylate, soaked in Tris-buffered saline-BSA, and incubated for 2 h at 25° C with ^{3}H-labeled mycoplasma suspended in RPMI 1640 containing 1% BSA and 25 mM Hepes, pH 7.3 (panel B), or sprayed with orcinol reagent to visualize glycolipids (panel A). Lanes 1, 1ug each of galactosylceramide (CMH), lactosylceramide doublet (CDH), trihexosylceramide (CTH), globoside GL4), and the gangliosides GM_3, GM_2, GM_1, GD_3, GD_{1a}, GD_{1b}, and GT_{1b}; lanes 2, 1ug of sulfatide; lanes 3, 0.5 ug of sulfatide; lanes 4, 0.25 ug of sulfatide; lanes 5, 0.12ug of sulfatide; lanes 6, 0.06 ug of sulfatide; lanes 7, 0.03 ug of sulfatide. See Table 1 for structures and abbreviations used.

determine the percent binding efficiency (Table 4). Greater than 99% of the *M. orale* bound to the gel containing the receptor analogue dextran sulfate, and no binding was detected to gels coupled with other polysaccharides or proteins. Similar binding efficiencies were obtained with the other mycoplasmas listed in Table 3 (data not shown). Thus, further optimization of this "receptor trap" to yield binding efficiencies approaching 100%, offers a novel means to process batch quantities of biofluids for removing contaminating mycoplasmas or viruses that have the same binding specificity. Optimization of these experiments is presently underway.

Table 2. Inhibition of *M. hominus* Binding to Sulfatide by Anionic Polysaccharides.

Inhibitor	I_{50} (μg/ml)[a]
Dextran sulfate (M_r 500,000)	5
Fucoidin	50
Dextran (M_r 500,000)	> 200
Heparin	> 200
Chrondroitin sulfate	> 200
Colominic acid	> 200

[a]Concentration giving 50% inhibition of *M. hominus* binding to 1 μg of sulfatide per well in a solid phase binding assay.

Table 3. Species from the Mycoplasma Group Which Bind Specifically to Sulfatide and are Inhibited by Dextran Sulfate.

Species[a]	Strain	Comment
M. pneumoniae	M129	Virulent, passage 4-6
M. hominus	ATCC 23114	Type strain
M. genitalium	ATCC 33530	Type strain
M. salivarium	ATCC 23064	Type strain;
M. orale	ATCC 23714	Type strain
M. hyorhinis	ATCC 17981	Neotype strain
M. arginini	ATCC 23838	Type strain
Acholeplasma laidlawii	ATCC 23206	Type strain
Ureaplasma urealyticum	ATCC 27618	Type strain

[a]Mycoplasmas were grown at 37° C in 5% CO_2 (except for *M. orale* and *M. salivarium* which were grown anaerobically) in modified Hayflick's medium containing arginine; *M. genitalium* was grown in *Spiroplasma* medium as recommended by the ATCC and *U. urealyticum* was grown in trypticase soy medium supplemented with 5% yeast extract and 0.1% urea.

Many Pathogenic Bacteria Bind to Ganglio- and Lacto-Series Glycolipids

We have previously reported that many pulmonary pathogenic bacteria, including *Pseudomonas aeruginosa*, *Haemophilus influenzae* and *Streptococcus pneumoniae*, specifically recognize the ganglio-series class of glycolipids (Krivan *et al.*, 1988 a and b) and that these bacteria required

Table 4. Binding Capacity and Recovery of *M. orale* from Immobilized Receptor Analogues.

Immobilized Ligand	*M. orale* Bound (cpm)	Effluent (cpm)	Binding Efficiency (%)[a]
Dextran Sulfate	1,071,350	9,900	99.1
Heparin	1,801	1,000,918	< 0
Ovalbumin	1,759	1,017	< 0
BSA	1,632	982,400	< 0

[a]Experimental binding efficiency was determined by using the equation (1-Effluent/Bound) x 100; in each experiment approximately 1 x 10^6 cpm of labeled mycoplasma was added to each column containing immobilized ligand.

terminal or internal GalNAcβ1-4Galβ1-4Glc sequences unsubstituted with sialyl residues (asialo gangliosides) for binding (see Table 6 for structures). Interestingly, *M. pneumoniae*, which binds to sulfatides (Krivan *et al.*, 1989), did not bind to asialo-GM_1 (Krivan *et al.*, 1988b). The biological relevance of these data was suggested by our finding that the receptor asialo-GM_1 was present in normal human lung tissue. This binding specificity, however, did not appear to be restricted to pathogens of the respiratory tract, as *Neisseria gonorrhoeae* also bound strongly to asialo-GM_1 (Stromberg *et al.*, 1988). Recently, we have analyzed the interaction of gonococci with a library of human glycolipids (Deal and Krivan, 1990) and found that in addition to binding to the ganglio-series glycolipids, asialo-GM_1 and asialo-GM_2, *N. gonorrhoeae* also specifically bound to the lacto-series class of glycolipids, which contain GlcNAcβ1-3Galβ-4Glc sequences. Binding to both classes of glycolipids was not dependent on pili, protein II, or the presence of lipooligosaccharide (Deal and Krivan, 1990). Because lacto-series structures are found in the lipooligosaccharides of many serovars of *N. gonorrhoeae* (Mandrell *et al.*, 1988), we proposed that the well-known phenomenon of gonococcal autoagglutination could be explained by the adhesin(s) of one gonococcus binding to GlcNAcβ1-3Galβ1-4Glc sequences in the lipooligosaccharide of another (Deal and Krivan, 1990).

The role of ganglio- and lacto-series glycolipids as adhesion receptors for other bacteria was further examined in light of this dual receptor specificity observed with *N. gonorrhoeae*. The glycolipid binding specificities of a variety of gram-positive and gram-negative bacteria were established by the bacterial overlay assay using high performance thin-layer chromatography (HPTLC), and are summarized in Table 5. For example, as shown by an autoradiogram (Fig. 2B) compared with an identical HPTLC plate visualized with orcinol reagent (Fig. 1A), a nonpiliated variant of *P. aeruginosa* PAO (kindly provided by William Paranchych, University of Alberta) bound avidly to asialo-GM_1 and its product obtained after β-galactosidase treatment, asialo-GM_2 (Fig. 2B, lane 2; Table 6). The organism also bound to purified human lacto-*N*-neotetraosylceramide (paragloboside) (Fig. 2B, lane 3), which occurs as a doublet due to differences in fatty acid content between the two glycolipid species (Fig. 2A, lane 3). Similar results were obtained with piliated *P. aeruginosa* (data not shown). As has been described for *N. gonorrhoeae*, both ganglio- and lacto-series glycolipids containing the minimum carbohydrate sequences GalNAcβ1-4Galβ1-4Glc and GlcNAcβ1-3Galβ1-4Glc, respectively, can support adhesion of *P. aeruginosa* and that binding to both structures is not dependent on pili. No binding was observed with other neutral and acidic glycolipids tested (Fig. 1B, lane 1), although in some experiments *P. aeruginosa* bound weakly to lactosylceramide (CDH) on HPTLC plates. However, this observation was not always reproducible (Table 6), and CDH

Table 5. Microorganisms Tested for Binding to ganglio-(GalNAcβ1-4Galβ1-4Glc...) and Lactoseries (GlcNAcβ1-3Galβ1-4Glc...) Glycolipids[a].

Microorganism	Ganglio-series	Lacto-series	Comment
Streptococcus pneumoniae	+++	++	ATCC 33400, Type strain
Streptococcus agalactiae	+++	+	ATCC 13813, Neotype
Neisseria gonorrhoeae	+++	+	$MS11_{mk}$, P^-,PII
Neisseria meningitidis	+++	+	ATCC 13077, Neotype
Haemophilus influenzae	+++	+	ATCC 9795, Type b
Pseudomonas aeruginosa	+++	+	CT3, CF patient
Pseudomonas cepacia	+++	+	ML1, CF patient
Pseudomonas maltophilia	+++	+	ATCC 13637, Type strain
Chlamydia trachomatis	++	-	E/UW-5/Cx, Serovar E
Chlamydia pneumoniae (TWAR)	++	-	ATCC VR-1310
Helicobacter pylori	++	-	ATCC 43504, Type strain

[a]Negative binding (-) indicates no binding to 2 μg of glycolipid and positive binding to less than 0.2μg (+++), 0.2 - 0.5 μg (++) and 1- 2 μg (+).

from a variety of sources, including human, did not demonstrate concentration-dependent binding (Fig. 3). Binding of *P. aeruginosa* to purified glycolipids immobilized on microtiter plates was examined to further define the binding specificity and to compare the relative avidities of the ganglio- and lacto-series receptors. As shown in Fig. 3, nonpiliated *P. aeruginosa* bound better to asialo-GM_1 than to galactosylparagloboside (or to paragloboside and its product obtained after β-galactosidase treatment, lacto-*N*-triaosylceramide (Table 6)), and not at all to lactosylceramide, consistent with the results obtained with the bacterial overlay assay (Fig. 2B). Similar glycolipid binding specificities were obtained with the other bacteria listed as positive for binding in Table 5.

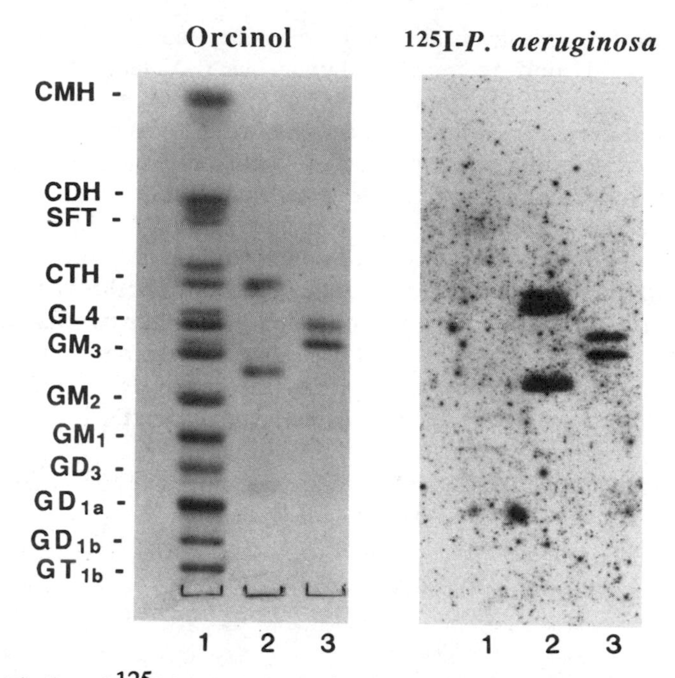

Figure 2: Binding of ^{125}I-labeled nonpiliated *P. aeruginosa* to glycolipids separated by thin layer chromatography. Panel A, glycolipids detected by orcinol reagent; panel B, autoradiogram of a chromatogram overlaid with radiolabled *P. aeruginosa* as described in the legend of Fig. 1. Lanes 1, 1 ug each of the standard glycolipids described in the legend of Fig. 1; lanes 2, 1ug each of asialo-GM_1 and asialo-GM_2; and lanes 3, 2 ug of paragloboside (doublet). See Table 6 for structures and abbreviations used.

Our finding that nonpiliated *P. aeruginosa* also recognizes the same glycolipids as piliated strains, suggests that adhesins other than pili may contribute to the pathogenesis of *P. aeruginosa*. The putative nonpiliated adhesins of the various bacteria listed in Table 5 that recognize asialo-GM_1 are probably also similar, at least with respect to the receptor binding domain of the adhesin molecule. For example, we have recently purified an adhesin protein from *H. influenzae* type b by receptor affinity chromatography using asialo-GM_1 (Fig. 4). Outer membrane proteins prepared from *H. influenzae* type b (Fig. 4, lane 1) and ^{35}S-labeled *H. influenzae* type b (Fig. 4, lane 2) were incubated with immobilized asialo-GM1 receptor or with globotetraosylceramide, a nonsense glycolipid that is not a receptor for the organism (Table 6). As shown by the autoradiogram in Fig. 4, a protein migrating between the major outer membrane proteins

P1 and P2 (arrow) was specifically eluted from immobilized asialo-GM1 and had an apparent molecular weight of about 41 kD (Fig. 4, lane 3). No visible ^{35}S-labeled protein, however, was eluted from the control, globotertraosylceramide (Fig. 4, lane 4). Similar results were obtained when whole cells of *H. influenzae* were surface iodinated with ^{125}I, sonicated and the radiolabeled soluble extract incubated with immobilized asialo-GM_1 (date not shown), suggesting that the adhesin is exposed on the surface of the organism. Interestingly, the asialo-GM_1 adhesin protein appears to be a minor outer membrane protein, as it is not readily visible by Coomassie blue staining (Fig. 4, lane 1). We have recently cloned the gene that codes for this protein and studies are currently underway to test the efficacy of this *H. influenzae* minor outer membrane as a possible adhesin vaccine in the infant rat model (J.E. Samuel and H.C. Krivan, manuscript in preparation).

Table 6. Glycolipids Tested for the Ability to Bind Many Bacteria on Thin-layer Chromatograms.

Glycolipid[a]	Structure	Binding[b]
Asialo-GM2	GalNAcβ1-4Galβ1-4Glcβ1-1Cer	+++
Asialo-GM1	Galβ1-3GalNAcβ1-4Galβ1-4Glcβ1-1Cer	+++
Lactotriaosylceramide	GlcNAcβ1-3Galβ1-4Glcβ1-1Cer	+
Paragloboside	Galβ1-4GlcNAcβ1-3Galβ1-4Glcβ1-1Cer	+
Glucosylceramide	Glcβ1-1Cer	-
Galactosylceramide (CMH)	Galβ1-1Cer	-
Lactosylceramide (CDH)	Galβ1-4Glcβ1-1Cer	+/-
Trihexosylceramide (CTH)	Galα1-4Galβ1-4Glcβ -1-Cer	-
Globotetraosylceramide (GL4)	GalNAcβ1-3Galα1-4Galβ1-4Glcβ1-1Cer	-
Forssman (Fors)	GalNAcα1-3GalNAcβ1-3Galα1-4Galβ1-4Glcβ-1-Cer	-
GM3	NeuAcα2-3Galβ1-4Glcβ1-1Cer	-
GM2	GalNAcβ1-4[NeuAcα2-3]Galβ1-4Glcβ1-1Cer	-
GM1	Galβ1-3GalNAcβ1-4[NeuAcα2-3]Galβ1-4Glcβ1-1Cer	-
Sialylparagloboside	NeuAcαGalβ1-4GlcNAcβ1-3Galβ1-4Glcβ1-1Cer	-
GD3	NeuAcα2-8NeuAcα2-3Galβ1-4Glcβ1-1Cer	-
GD1a	NeuAcα2-3Galβ1-3GalNAcβ1-4[NeuAcα2-3Galβ1-4Glcβ1-1Cer	-
GD1b	Galβ1-3GalNAcβ1-4[NeuAcα2-8NeuAcα2-3]Galβ1-4Glcβ1-1Cer	-
GT1b	NeuAcα2-3Galβ1-3GalNAcβ1-4[NeuAcα2-8NeuAcα2-3Galβ1-4Glcβ1-1Cer	-

[a]Trivial names and structures are represented according to recommendations in Reference 14 and references cited therein; cer, ceramide; CMH, ceramide monohexoside; CDH, ceramide dihexoside (lactosylceramide); CTH, ceramide trihexoside (globotriaosylceramide); GL4, globotetraosylceramide.

[b]Negative binding (-) indicates no binding to 2 µg of glycolipid and positive binding to less than 0.4 µg (+++), 0.8-1 µg (+), and greater than 2 µg (+/-).

Summary and Perspective

We have shown that the carbohydrate sequences of glycolipids encode very specific binding domains for microbial attachment. Many mycoplasmas bind tightly to sulfatides, ubiquitous glycolipids (Hakomori, 1987) found in many human and animal cells, and this specific binding is strongly inhibited by dextran sulfate. They do not bind to asialo-GM_1 (Table 1). By immobilizing dextran sulfate on an insoluble support, this polyvalent receptor analogue can be exploited as an efficient means to remove mycoplasmas that contaminate biofluids, such as fetal calf serum or tissue culture media. Sulfatide-mediated adhesion, however, is more than just biotechnology, as it may also be important in mycoplasma pathogenicity to guarantee intimate contact of the parasite with the host cell membrane to satisfy its strict

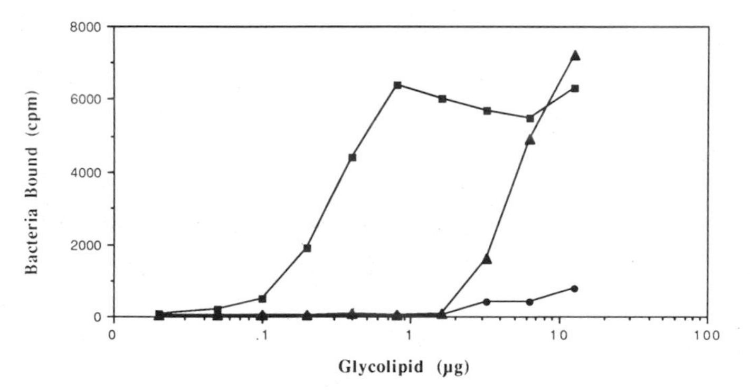

Figure 3: Binding of nonpiliated *P. aeruginosa* to purified glycolipids adsorbed in microtiter wells. Lipids in 25 ul of methanol containing 0.1 ug each of the auxiliary lipids cholesterol and phosphatidylcholine were evaporated in flat-bottom wells of polyvinylchoride microtiter plates. The wells were blocked with 1% albumin overnight at 4°C washed twice with saline, and incubated at 25°C with 25 ul of ^{125}I-labeled *P. aeruginosa* (approximately 10^5 cpm). After 2 h, the wells were washed 5 times with cold saline, cut from the plate, and bound radioactivity was quantified in a gamma counter. *P. aeruginosa* binding was determined in Tris-BSA for asialo-GM_1 (■), galactosylparagloboside (▲), and lactosylceramide (●).

nutritional requirements. Our finding that many species within the *Mycoplasmatales* share a common glycolipid receptor suggests that a common adhesin is present in all of these microorganisms. The idea of a common or universal receptor and its corresponding universal adhesin or binding domain for certain groups of microorganisms could explain the results described in this paper. The mycoplasma group binds to sulfatide and it follows that their sulfatide-binding adhesins may be very similar. Likewise, the bacteria listed in Table 5 bind strongly, for example, to terminal or internal GalNAcβ1-4Galβ1-4Glc sequences unsubstituted with sialyl residues (Table 6), again, suggesting that a common adhesin or universal binding domain on the surface of these bacteria exists for this receptor. Interestingly, several of the bacteria listed in Table 5 are known to produce neuraminidases (sialidases), enzymes that can hydrolyze sialyl residues of gangliosides and glycoproteins. Why do these bacteria produce such an enzyme, which presumably is a metabolically expensive thing to do? The answer may be that the ability of a microorganism to produce neuraminidase may directly increase its virulence, as the microbial enzyme would unmask cryptic host cell surface receptors and create additional asialo-ganglioside binding sites for colonization. Many viruses, including influenza virus, also express neuraminidases. A possible explanation as to why influenza virus and some of the opportunistic pathogens listed in Table 5 might interact in the pathogenesis of secondary pneumonia is that the viral neuraminidase may desialylate carbohydrate sequences on the cells lining the respiratory tract and increase the number of structures containing

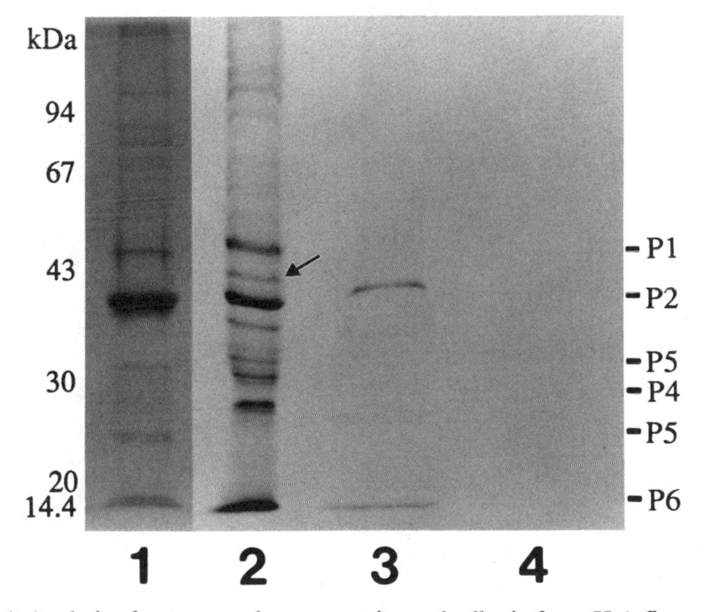

Figure 4: Analysis of outer membrane proteins and adhesin from *H. influenzae* type b (ATCC 9795) by SDS-PAGE and autoradiography. Lanes: 1, outer membrane protein preparation visualized by Coomassie blue stain; 2, autoradiogram of ^{35}S-labeled outer membrane protein (arrow indicates adhesin); 3, autoradiogram of ^{35}S-labeled protein eluted from immobilized asialo-GM$_1$; and 4, autoradiogram of ^{35}S-labeled protein eluted from immobilized globotetraosylceramide. Molecular weight standards (left) and outer membrane protein nomenclature (right) are as indicated, i.e., P1, is outer membrane protein P1 (Granoff and Munson, 1986).

unsubstituted Galβ1-3GalNAcβ1-4Galβ1-4Glc and/or Galβ1-4GlcNAcβ1-3Galβ1-4Glc residues. Thus, the flu virus probably paves the way for increased microbial colonization.

The use of one receptor by many species of pathogenic bacteria may imply an evolutionary convergence by certain groups of microorganisms. The adhesin protein isolated from *H. influenzae* type b (Fig. 4, lane 3) has been compared with the adhesins from *C. trachomatis*, *H. pylori*, *P. aeruginosa* and *S. pneumoniae*, and is surprisingly similar (H.C. Krivan and C.A. Lingwood, manuscript in preparation). While most bacteria exhibit serotype diversity, the conservation of a universal adhesin or binding domain, even among different genera, would ensure that the initial process of adhesion to eucaryotic cell surface receptors, which are also conserved, would occur. We are actively pursuing this line of reasoning, and are exploiting the receptor binding specificities of several microorganisms to further study the structural and molecular aspects of these common microbial adhesin proteins. Whether an adhesin vaccine or "pan vaccine" could be developed based on these findings, which could protect humans from several bacterial diseases, remains to be seen.

References

Beachey, E. H., 1981, Bacterial adherence: adhesin-receptor interactions mediating the attachment of bacteria to mucosal surfaces, *J. Infect. Dis.* 143: 325-345.

Jones, G.W. and R.E. Isaacson, 1985, Proteinaceous bacterial adhesins and their receptors, *CRC Crit. Rev. Microbiol.* 10: 229-260.

Karlsson, K-A., 1989, Animal glycosphingolipids as membrane attachment sites for bacteria, *Annu. Rev. Biochem.* 58: 309-350.

McGarrity, G. J., D. Meredith, D. Gruber, and M. McCall, 1988, In: Mycoplasma Infection of Cell Cultures. Ed: G.J. McGarrity, D.G. Murphy, ad W.W. Nichols, Plenum, New York, p. 243-334.

Krivan, H.C., L.D. Olson, M.F. Barile, V. Ginsburg, and D.D. Roberts, 1989, Adhesion of *Mycoplasma pneumoniae* to sulfated glycolipids and inhibition by dextran sulfate, *J. Biol. Chem.* 264: 9283-9288.

Lingwood, C.A., P.A. Quinn, S. Wilansky, A. Nutikka, H.L. Ruhnke, and R.B. Miller, 1990, Common sulfoglycolipid receptor for mycoplasmas involved in animal and human infertility, *Biol. Reproduction* 43: 694-697.

Krivan, H.C., V. Ginsburg, and D.D. Roberts, 1988, *Psuedomaonas aeruginosa* and *Psuedomonas cepacia* isolated from cystic fibrosis patients bind specifically to gangliotetraosylceramide (asialo-GM_1 and) gangliotriaosylceramide (asialo-GM_2), *Arch. Biochem. Biophys.* 260: 493-495.

Krivan, H.C., D.D. Roberts, and V. Ginsburg, 1988b, Many pulmonary pathogenic bacteria bind specifically to the carbohydrate sequence

GalNAcβGal found in some glycolipids, *Proc. Natl. Acad. Sci.* 85: 6157-6161.

Stromberg, N., C. Deal, G. Nyberg, S. Normark, M. So, and K-A. Karlsson, 1988, Identification of carbohydrate structures that are possible receptors for *Neisseria gonorrhoeae*, *Proc. Natl. Acad. Sci.* 85: 4902-4906.

Deal, C.D. and H.C. Krivan, 1990, Lacto- and ganglio-series glycolipids are adhesion receptors for *Neisseria gonorrhoeae*, *J. Biol. Chem.* 265: 12774-12777.

Mandrell, R.E., J.M. Griffiss, and B.A. Macher, 1988, Lipooligosaccharides (LOS) of *Neisseria gonorrhoeae* and *Neisseria meningitidis* have components that are immunologically similar to precursors of human blood group antigens, *J. Exp. Med.* 168:107-126.

Granoff, D.M. and R.S. Munson, Jr., 1986, Prospects for prevention of *Haemophilus influenzae* type b disease by immunization, *J. Infect. Dis.* 153: 448-461.

Hakomori, S-i., 1987, Chemistry of glycosphingolipids In: Handbook of Lipid Research 3, Sphingolipid Biochemistry. Ed: J.N. Kanfer and S-i. Hakomori, Plenum Press, New York, p. 1-150.

IUPAC-IUB Joint Commission on Biochemical Nomenclature, *Eur. J. Biochem.* 159: 16, 1986.

The Integrins: A General Overview

Jonathan M. Edelman and Clayton A. Buck

Introduction

The integrins are a family of cell surface molecules involved in several important events including cell substratum adhesion, cell-cell adhesion, cell migration, cell differentiation and signal transduction. They include receptors for extracellular matrix molecules such as fibronectin, laminin and collagen (Hynes, 1987; Buck and Horwitz, 1987b; Ruoslahti and Pierschbacher, 1987; Akiyama *et al.*, 1990) as well as receptors that are involved in lymphocyte adhesion to one another and to the vascular endothelium (Springer, 1990; Nakamura *et al.*, 1990; Albelda and Buck, 1990; Elices *et al.*, 1990). Their importance to neural crest cell migration (Bronner-Fraser, 1985; Bronner-Fraser, 1986), gastrulation (Boucaut *et al.*, 1984), and neurogenesis (Reichardt and Tomaselli, 1991) during embryonic development has been well documented. They also appear to function as mediators of signals from the extracellular environment. For example, antibodies that react with integrins can interfere with muscle development (Menko and Boettiger, 1987) as well as stimulate the secretion of proteolytic enzymes (Werb *et al.*, 1989). Integrins also serve as accessory molecules during the stimulation of lymphocytes (Shimizu *et al.*, 1990b; Shimizu *et al.*, 1990a). Finally, they serve as the binding sites for certain pathogenic bacteria and fungi (Leong *et al.*, 1990; Isberg and Leong, 1990; Bullock and Wright, 1987). Thus, although integrins were first recognized as adhesion promoting molecules, they clearly have other important biological functions.

Integrins: General Properties of the Receptor Family

Structurally, integrins are noncovalently associated transmembrane heterodimers consisting of an α and a β subunit. Only the assembled heterodimer found on the cell surface is fully functional. The integrin family has been organized into three major subfamilies on the basis of a common β subunit (Fig. 1). The largest subfamily is the β_1 or VLA subfamily (Hynes, 1987; Hemler, 1990; Albelda and Buck, 1990). This subfamily includes the integrins that function as extracellular matrix receptors such as the fibronectin receptor ($\alpha_5\beta_1$) as well as several receptors for laminin, and collagen. At least three members of this subfamily, $\alpha_2\beta_1$, $\alpha_3\beta_1$ and $\alpha_4\beta_1$ have been implicated in cell-cell adhesion (Carter *et al.*, 1990b;

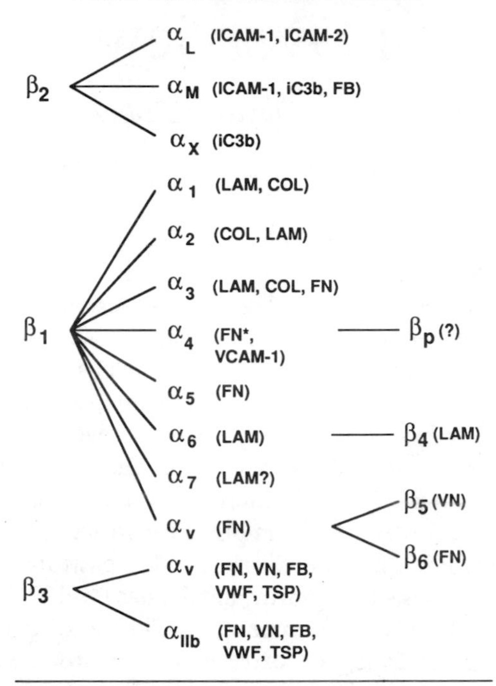

Figure 1. Integrin Heterodimers and their Ligands. Known integrin heterodimer combinations are indicated by lines connecting α and β subunits. Ligands for each heterodimer combination are indicated in parenthesis. ICAM-1=intercellular adhesion molecule 1; ICAM2=intercellular adhesion molecule 2; iC3b=C3 complement fragment; FB=fibrinogen; LAM=laminin; COL=collagens; FN=fibronectin; FN^*=alternatively spliced fibronectin; VCAM-1=vascular cell adhesion molecule 1; VN=vitronectin; VWF=vonWillebrand factor; TSP=thrombospondin.

Larjava *et al.*, 1990; Elices *et al.*, 1990; Holzman and Weissman, 1989; Campanero *et al.*, 1990)

The β_2 subfamily or Leu-CAM's include the receptors found primarily on white cells that are involved in lymphocyte-lymphocyte interactions as well as interactions between leukocytes and vascular endothelium, and phagocytosis of complement opsonized pathogens

(Springer, 1990). In addition, β_2 integrins have been reported on several carcinomas (Blanchard *et al.*, 1990; Feldman *et al.*, 1991). This subfamily is particularly important in inflammatory and hyperimmune responses (Albelda and Buck, 1990). Individuals carrying a mutation coding for a functionally compromised β_2 subunit have been identified (Anderson and Springer, 1987; Arnaout *et al.*, 1990; Wardlaw *et al.*, 1990). These individuals suffer from a disease known as leukocyte adhesion deficiency (LAD) in which their white cells are unable to participate in transendothelial migration. Such patients suffer from chronic nonsuppurartive and eventually fatal infections due to the inability of their neutrophils to move to areas of inflammation and to phagocytose opsonized particles. The ligands for this subfamily of integrins include the iC3b fragment of complement, fibrinogen and members of the immunoglobulin superfamily of molecules such as ICAM-1 and ICAM-2 (Springer, 1990; Wawryk *et al.*, 1989).

The third subfamily is the β_3 integrins also known as cytoadhesins (Ginsberg *et al.*, 1988; Phillips *et al.*, 1988). These integrins include the major platelet receptor, gpIIb/IIIa that is active in thrombus formation and which, if defective, as in the case of Glanzmann's thrombesthenia, results in a bleeding disorder. This family also includes receptors for fibrinogen, von Willebrand factor, thrombospondin, and vitronectin.

Several general properties of integrins are evident in Figure 1. For example, it is the combination of a particular α and β subunit that imparts the specific binding properties to each heterodimer. Thus, $\alpha_5\beta_1$ is a fibronectin receptor (Pytela *et al.*, 1985) while $\alpha_6\beta_1$ is a laminin receptor (Sonnenberg *et al.*, 1988a). For the most part, a single β subunit can combine with several α subunits; however, there are instances in which a single α subunit combines with more than one β subunit, as with α_v, α_4, and α_6 (Hemler *et al.*, 1989; Sonnenberg and Linders, 1990; Sonnenberg *et al.*, 1988b; Smith *et al.*, 1990). In each case, this change in β subunit association is accompanied by a change in ligand specificity.

While certain integrins interact with only a single ligand, others clearly interact with several different molecules. An example of this is $\alpha_3\beta_1$, which has been variously designated as a laminin, collagen or fibronectin receptor (Elices *et al.*, 1991). The reasons for the ambiguity in determining the specificity of this integrin may be that its ligand binding is somehow dictated by the cell in which it is expressed. This integrin is also engaged in cell-cell interactions that may involve as yet uncharacterized ligands (Peltonen *et al.*, 1989; Larjava *et al.*, 1990; Carter *et al.*, 1990b).

There are multiple integrin receptors for specific ligands. For example, there are now seven different integrins that interact with fibronectin and at least six that bind laminin. In the case of fibronectin two different regions of this large and complicated extracellular matrix molecule have been identified as integrin ligands. The integrin $\alpha_5\beta_1$ binds to the RGD sequence located in the cell binding domain of fibronectin (Ruoslahti and Pierschbacher, 1987) whereas $\alpha_4\beta_1$ binds to the LDV peptide in the alternatively spliced region of the molecule (Wayner *et al.*, 1989; Guan and Hynes, 1990). Similarly, ligands such as collagen and laminin are large molecules with structurally distinct regions. These regions may serve as binding sites for different integrins thereby resulting in the existence of multiple integrins for a particular ligand.

Integrins as Transmembrane Mediators of Adhesive Events

The fact that monoclonal antibodies against specific integrins block adhesive events argues that integrins are somehow involved in mediating cell-matrix or cell-cell adhesion (Buck and Horwitz, 1987b). This contention is further strengthened by the fact that integrins can be found concentrated in adhesive organelles such as focal contacts (Chen *et al.*, 1986; Carter *et al.*, 1990a; Damsky *et al.*, 1985) and hemidesmosomes (Stepp *et al.*, 1990) both of which are also sites where cytoskeletal molecules and extracellular matrix molecules are found (Burridge *et al.*, 1988; Chen *et al.*, 1986). Furthermore, purified integrins bind directly to their ligands in a saturable, cation dependent and specific manner (Buck and Horwitz, 1987a; Ruoslahti and Pierschbacher, 1987; Ruoslahti, 1991). Where the specific sites of these interactions are known, as in the case of the cell binding domain of fibronectin, the binding of the receptor to the ligand can be inhibited by specific agents such as the RGD peptide (Yamada and Kennedy, 1984; Pierschbacher and Ruoslahti, 1984) or biologically active monoclonal antibodies (Neff *et al.*, 1982).

Integrins also interact directly with the cytoskeletal associated molecules talin (Buck and Horwitz, 1987a) and α-actinin (Otey *et al.*, 1990). This binding occurs exclusively on that portion of the integrin that extends into the cytoplasm of the cell. The picture that emerges is one in which the binding of integrins to extracellular ligands results in the clustering of these receptors at the site of ligand interaction and triggers the organization of cytoskeletal elements required for the formation of adhesive structures. The precise mechanism by which this is accomplished and regulated is not known. There is some suggestion however, that subunit modification such as phosphorylation may serve to control this process (Tapley *et al.*, 1989; Horvath *et al.*, 1990; Buck and Horwitz, 1987a). The structure of a "generic" integrin molecule is shown in Figure 2. This structure is based upon cDNA sequence data, electron microscopy of purified integrins (Nermut *et al.*, 1988) and biochemical determination of precise cysteine pairing from careful studies of the subunits of the platelet integrin $\alpha_{IIb}\beta_{IIIa}$ (Beer and Coller, 1989; Calvete *et al.*, 1991). Given this structure and what is known about integrin function, there are certain functional domains which remain to be elucidated. These include the ligand binding domain, the subunit association domain and the cytoskeletal organizing domain. Some progress has been made in the identification of two of these

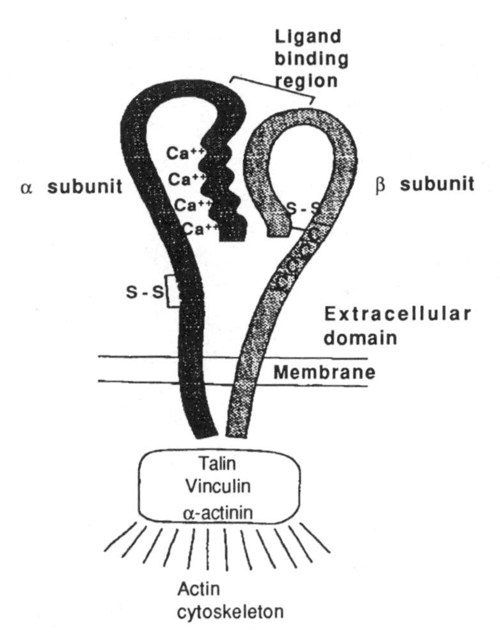

Figure 2. Schematic Structure of an Integrin.

functional domains. Peptide cross linking studies indicate that the binding of two ligands, fibronectin and vitronectin, occurs in the region designated in Fig. 2 (D'Souza *et al.*, 1988; D'Souza *et al.*, 1991; Smith and Cheresh, 1988). This domain is conserved in most integrins suggesting that it will turn out to be the ligand binding site on many of these receptors.

The function of the cytoplasmic domain in cytoskeletal organization and in targeting integrins to adhesive structures such as focal contacts has been determined by expressing cDNA's coding for β subunits with altered cytoplasmic domains (Solowska *et al.*, 1989; Hayashi *et al.*, 1990; Marcantonio *et al.*, 1990). If certain portions of the β subunit cytoplasmic domain are deleted, integrins carrying this altered β subunit are not found in focal contacts and do not take part in adhesive interactions. Similar experiments have not been reported for the cytoplasmic domain of α subunits, so the role the cytoplasmic portion of α subunits in cytoskeleton organization is presently undefined.

Chimeric β subunits have been constructed in which portions of the β_1 subunit are replaced with homologous regions of the β_3 subunit in an attempt to identify regions of the molecule that are required for a subunit selection. These experiments, like the cross-linking experiments, suggest that the ligand binding domain as well as the regions of the β subunit responsible for a subunit selection lie in the extracellular globular 40% of the subunit (Edelman, Shih and Buck, unpublished observations).

There are clearly other domains of interest on integrins. Some are revealed following ligand binding (Frelinger *et al.*, 1990), and others appear to be involved in regulating integrin function (Neugebauer and Reichardt, 1991). These latter sites have been identified as epitopes on the β_1 subunit of avian integrins which are involved in modifying integrin binding activity. When a monoclonal antibody binds to this site on the subunit, the interaction of the receptor with vitronectin is diminished while the binding to laminin and collagen is enhanced (Neugebauer and Reichardt, 1991). The existence of such epitopes suggests that mechanisms exist for modifying the affinity of integrins for various ligands. This may be an important mechanism for modulating the activity of cells within tissues or for facilitating cell motility.

Expression of Integrins by Normal and Diseased Tissue

The integrin repertoire of various cells and tissues has been investigated using antibodies specific for integrin subunits. Integrins expressed in tissues have been identified by immunohistochemistry, while immunoprecipitation and immunoblotting have been used to identify integrins present in nonionic detergent extracts from cultured cells. In both cases, it has been established that cells express multiple integrins (Albelda *et al.*, 1989; Albelda *et al.*, 1991b). However, frequently there are differences between integrins expressed by a particular cell type in culture and those found on the same cells *in situ*. This is particularly evident in the case of endothelial cells (Albelda *et al.*, 1989). All endothelial cells in culture express the collagen/laminin receptors as well as receptors for fibronectin and vitronectin; however, *in situ* immunohistochemical staining of vasculature shows that the expression of receptors for fibronectin and vitronectin is a property of large vessel endothelium and not of small vessel endothelium (Table 1 and S. Albelda and C. Buck, unpublished observations).

Table 1. *In Situ* Integrin Expression by Normal Tissue

	INTEGRIN SUBUNIT							
	Collagen-Laminin Receptors				Fibronectin-Fibrinogen Receptors			
TISSUE	α_1	α_2	α_3	α_6	α_4	α_5	α_v	β_3
KERATINOCYTES	*	●	●	●	○	○	●	○
BRONCHIAL EPITHELIUM	*†	●	●	●	○	○	●	○
DERMAL EPITHELIUM	*†	●	●	●	○	○	●	○
LARGE VESSEL ENDOTHELIUM	●†	●†	●	●	○	●	●	●
CAPILLARY ENDOTHELIUM	●†	●	●	●	○	○†	○†	○†
AIRWAY SMOOTH MUSCLE	●	●	●	○	○	○†	●	○
VASCULAR SMOOTH MUSCLE	○	●	●	○	○	○†	●	○†
MELANOCYTES (NEVUS)	*†	●	●	●	○†	○†	●	○†

● = Readily detectable   * = Trace   ○ = Not detectable
† = Different from same cells in culture

A summary of integrin expression by various normal adult human tissues is shown in Table 1. As a general rule, it is the laminin/collagen receptors that are most widely expressed *in situ*. The fibronectin/fibrinogen receptors are expressed by a more restricted group of tissues (Buck *et al.*, 1990). Except for large vessel endothelium and certain glandular tissues, the classical fibronectin receptor $\alpha_5\beta_1$ is rarely seen in normal adult tissue. An *in situ* evaluation of integrin expression by tumors of epithelial origin reveals no consistent change of pattern between the normal and transformed cells,

Table 2. *In Situ* Integrin Expression by Various Tumors

	INTEGRIN SUBUNIT							
	Collagen-Laminin Receptors				Fibronectin-Fibrinogen Receptors			
TISSUE	α_1	α_2	α_3	α_6	α_4	α_5	α_v	β_3
MELANOCYTES (NEVUS)	✱†	●	●	●	○†	○†	●	○†
MELANOMA (RGP)	○	●	●	●	✱	○	●	○
MELANOMA (VGP OR MET)	○	●	●	●	✱	✱	●	●
LARGE CELL CARCINOMA 1+	○	●	✱	○	○	✱	●	✱
LARGE CELL CARCINOMA 2+	○	○	○	●	○	○	○	○
ADENOCARCINOMA 1+	○	●	●	○	○	○	✱	○
ADENOCARCINOMA 2+	✱	●	✱	○	○	○	●	○
SQUAMOUS CELL CARCINOMA 1+	✱	○	✱	○	○	○	✱	○
SQUAMOUS CELL CARCINOMA 2+	✱	●	●	●	○	✱	●	✱

● = Readily detectable ✱ = Trace ○ = Not detectable
† = Different from same cells in culture + = Tumors of the lung

(Table 2) regardless of the tissue involved, i.e. lung, breast or colon. For the most part, there appears to be a "down regulation" of integrin expression in carcinomas as noted by a decrease in the intensity of immune staining. However, for carcinomas in general, no predictable correlation between specific tumor stages and integrin expression has emerged. In benign adenomas and highly differentiated adenocarcinomas, integrin expression more nearly reflects that of the tissue of origin (Pignatelli and Bodmer, 1990; Zutter *et al.*, 1990; Koretz *et al.*, 1991). Those transformed cells still in contact with a basement membrane show no deviation from the normal pattern of integrin expression. However, more aggressive tumors, such as small cell carcinoma of the lung, display a marked down-regulation of integrin expression to the point that in some instances, β_1 integrins are nearly undetectable (Buck *et al.*, 1990; Albelda *et al.*, 1991b; Feldman *et al.*, 1991).

The one exception noted so far to the general pattern of integrin down regulation in tumors is found in human melanomas. Melanomas are tumors that develop in well defined stages. The first stage is the radial growth phase (RGP) characterized by a highly invasive tumor that spreads in a radial fashion into the dermis and epidermis. The integrins expressed by tumors in this phase are identical to those found in nonmalignant melanocytes and cells within benign nevi. Without treatment, melanomas eventually progress to a highly aggressive, phase which is always metastatic, called the vertical growth phase (VGP). Tumors entering this phase as well as those that have metastasized always express the β_3 subunit not found in normal melanocytes or RGP melanomas (Albelda *et al.*, 1991a; McGregor *et al.*, 1989). VGP melanomas frequently, but not always, also express the α_4 integrin subunit. This unusual increase in integrin expression correlates with the metastatic potential of the tumor. Whether or not the change in integrin expression facilitates the metastatic phenotype has yet to be determined. However, monoclonal antibodies directed against this integrin have been shown to inhibit melanoma growth *in vivo* (Boukerche *et al.*, 1989)

Integrins are expressed in consistent and reliable patterns in adult tissue. Why a cell is expressing more than one integrin is not clear. Perhaps the answer lies in the multiple functions of integrins, or in unknown and as yet to be discovered ligand interactions. Their importance as signal transducing molecules may determine the need for cells to express various integrins. It is not yet known at what point in development the adult integrin repertoire of any given tissue is established; however, there appears to be developmental regulation of integrin expression (Neugebauer and Reichardt, 1991; Cohen *et al.*, 1989; Ruoslahti and Pierschbacher, 1987; Patel and Lodish, 1985). This is not surprising since integrin expression can be influenced by cytokines such as TFG-β (Massague, 1990) that are known to influence cell behavior during embryonic development. The fact that integrin expression changes with diseased states further suggests that there must be a rather plastic program of integrin expression that may change with the various cellular activities associated with wound healing, angiogenesis, neurite extension, cell migration, etc. It is clear that different tissues express different integrins; that integrin expression can be regulated (Inghirami *et al.*, 1990; Massague, 1990; Fingerman and Hemler, 1988; Nicholson and Watt, 1991); that integrin expression changes during development and during various pathogenic processes; and that the major integrins found in adult tissue are those involved with collagen/laminin adhesion.

Many bacteria are recognized by phagocytic cells once they have been opsonized with complement via the $\alpha_M\beta_2$ integrin (also known as CR3). However several organisms are capable of binding to phagocytic cells directly, without first being opsonized. β_2 integrins found on neutrophils are able to bind to lipopolysaccharide lipid IVa, found on rough *E. coli*

Table 3. Pathogen Adherence to Eukaryotic Cell Integrins

Pathogen	Surface Protein	Target Cell	Integrin	RGD[f]	Reference
B. pertussis	Pertactin	CHO[d]	unknown	yes	(Leininger et al., 1991)
"	FHA[a]	Macrophage	αMβ2	yes	(Relman et al., 1990)
E. coli (rough)	LPS[b] lipid IVa	PMN[e]	β2(unspecified)	no	(Wright et al., 1989)
H. capsulatum	unknown	M. phi[d]	all β2	unknown	(Bullock and Wright, 1987)
L. mexicana	gp63	Macrophage	αMβ2	yes	(Russell and Wright, 1988)
"	LPG[c]		αMβ2, αXβ2	no	(Rohana et al., 1991)
Y. pseudo-	Invasin	HEp-2[d]	α5β1	unknown	(Leong et al., 1990; Isberg
tuberculosis	"	EJ[d]	α3β1, α5β1	unknown	and Leong, 1990)
"	"	K527[d]	α5β1	unknown	
"	"	HPB MLT[d]	α4β1	unknown	
"	"	Platelet	α6β1	no	

a Filamentous hemagglutinin; *b* Lipopolysaccharide; *c* Lipophosphoglycan; *d* Cell culture line; *e* Polymorphonuclear leukocyte; *f* RGD involvement indicates that the adhesive interaction can be inhibited in the presence of soluble RGD containing peptides.

(Wright *et al.*, 1989). Attachment of *Histoplasma capsulatum* to the cell surface is apparently mediated by all three β_2 integrins (Bullock and Wright, 1987). The fungal ligand(s) for this process are unknown. The macrophage integrin, $\alpha_M\beta_2$, is also capable of binding directly to several bacterial surface proteins. For example, two proteins found on *Leishmania mexicana* promastigotes, gp63 and lipophosphoglycan (LPG), can serve as ligands for $\alpha_M\beta_2$ (Rohana-Talamas *et al.*, 1991; Russell and Wright, 1988). In addition, LPG can bind to $\alpha_X\beta_2$ but not to $\alpha_L\beta_2$ (Rohana-Talamas *et al.*, 1991). A second surface protein found on *B. pertussis*, filamentous hemagglutinin (FHA), serves as a ligand for macrophage $\alpha_M\beta_2$ (Relman *et al.*, 1990). FHA can bind directly to $\alpha_M\beta_2$ in a manner that is inhibitable by RGD peptides.

Future Directions

Integrins are one of the newest family of receptors to be identified. Their role in both normal and disease processes will continue to be an area of major activity. The therapeutic manipulation of integrin function has already proven beneficial in promoting wound healing and in preventing tissue damage in ischemia reperfusion injury (Vedder *et al.*, 1988; Carlos and Harlan, 1990), heart lesions (Simpson *et al.*, 1988) and in meningitis (Tuomanen *et al.*, 1989). Inhibition of β_2 integrin function has also been shown to decrease the inflammatory response due to lipopolysaccharide stimulation of endothelium (Rosen and Gordon, 1990). Work in this area will continue as more is discovered about the role of integrins in various disease processes.

At a more basic level, there is much remaining to be learned about the molecules themselves. Work is just beginning on the identification of specific functional domains, and little is known concerning modifications to the molecules that can result in altered ligand specificity or avidity. As pointed out earlier, integrins are most certainly involved in second message generation required in enzyme induction and cellular differentiation. Little is known about the regulation of integrin expression at the genetic level, and the mechanisms regulating integrin expression on the cells' surface. This area of investigation will be extremely important with respect to the role of integrins in regulating cell-cell and cell matrix interactions during embryonic development. The mechanisms whereby integrins serve as receptors for bacterial invasion have yet to be worked out. Finally, there are already over 18 known integrins, and more are being discovered almost monthly. Much future activity will be focused on the discovery of new integrins or of integrins with modified structure. While there has been a great deal of progress toward understanding the molecular basis of integrin function, there is certainly much to be learned in the future.

Acknowledgements

This work supported by grants CA 19144; CA 10815 and HL 39023 to C. Buck.

References

Akiyama, S.K., Nagata, K., and Yamada, K., 1990, Cell surface receptors for extracellular matrix components, *Biochim. Biophys. Acta* 103, 91-110.

Albelda, S.M. and Buck, C.A., 1990, Integrins and other cell adhesion molecules, *FASEB J.* 4, 2868-2880.

Albelda, S.M., Daise, M., Levine, E.N., and Buck, C.A., 1989, Identification and characterization of cell-substratum adhesion receptors on cultured human adult large vessel endothelial cells, *J. Clin. Invest.* 83, 1992-2002.

Albelda, S.M., Mette, S.A., Elder, D.E., Stewart, R.M., Damjanovich, L., Herlyn, M., and Buck, C.A., 1991a, Integrin distribution in Malignant Melanoma: Association of the β_3 subunit with tumor progression, *Cancer Res.* 50, 6757-6764.

Albelda, S.M., Solowska, J., Edelman, J.M., Damjanovich, L., and Buck, C.A., 1991b, The role of integrins in development: Structure, function, and tissue specific expression. In: The Avian Model in Developmental Biology: From Organism to Genes. N. LeDouarin, F. Dieterlen-Lievre, and J. Smith, eds. (Paris: Edition du CNRS), pp. 261-280.

Anderson, D.C. and Springer, T.A., 1987, Leukocyte adhesion deficiency: an inherited defect in the Mac-1, LFA-1 and p150,95 glycoproteins, *Annu. Rev. Med.* 38, 175.

Arnaout, M.A., Dana, N., Gupta, S.K., Tenen, D.G., and Fathallah, D.M., 1990, Point mutations impairing cell surface expression of the common β subunit (CD18) in a patient with leukocyte adhesion molecule (Leu-Cam) deficiency, *J. Clin. Invest.* 85, 977.

Beer, J. and Coller, B.D., 1989, Evidence that platelet glycoprotein IIIa has a large disulfide-bonded loop that is susceptible to proteolytic cleavage, *J. Biol. Chem.* 264, 17564-17573.

Blanchard, D.K., Hall, R.E., and Djeu, J.Y., 1990, Role of CD18 in lymphokine activated killer (LAK) cell-mediated lysis of human monocytes: Comparison with other LAK targets, *Int. J. Cancer* 45, 312-319.

Boucaut, J.C., Darribere, T., Poole, T.J., Aoyama, H., Yamada, K.M., and Thiery, J.P., 1984, Biological active synthetic peptides as probes of embryonic development: a competitive peptide inhibitor of fibronectin function inhibits gastrulation in amphibian embryos and neural crest cell migration in the avian embryo, *J. Cell Biol.* 99, 1822-1830.

Boukerche, H., Berthier-Vergnes, O., Bailly, M., Dore, J.F., Leung, L.K., and McGregor, J.L., 1989, A monoclonal antibody (LYP18) directed against the blood platelet glycoprotein IIb/IIIa complex inhibits human melanoma growth *in vivo*, *Blood* 74, 909-912.

Bronner-Fraser, M., 1985, Alterations in neural crest migration by a monoclonal antibody that affects cell adhesion, *J. Cell Biol.* 101, 610-617.

Bronner-Fraser, M. 1986, An antibody to a receptor for fibronectin and laminin perturbs cranial neural crest development *in vivo*, *Dev. Biol.* 117, 528-536.

Buck, C., Albelda, S., Damjanovich, L., Edelman, J., Shih, D.-T., and Solowska, J. 1990, Immunohistochemical and molecular analysis of β_1 and β_3 integrins, *Cell Differ. & Develop.* 32, 189-202.

Buck, C.A. and Horwitz, A.F. 1987a, Integrin, a transmembrane glycoprotein complex mediating cell-substratum adhesion, *J. Cell Sci.* 8, 231-250.

Buck, C.A. and Horwitz, A.F. 1987b, Cell surface receptors for extracellular matrix molecules, *Annu. Rev. Cell Biol.* 3, 179-205.

Bullock, W.E. and Wright, S.D. 1987, Role of the adherence-promoting receptors, CR3, LFA-1, and p150,95, in binding of *Histoplasma capsulatum* by human macrophages, *J. Exp. Med.* 165, 195-210.

Burridge, K., Fath, K., Kelly, T., Nuckolis, B., and Turner, C. 1988, Focal adhesions: Transmembrane junctions between the extracellular matrix and the cytoskeleton, *Ann. Rev. Cell Biol.* 4, 487-525.

Calvete, J.J., Henschen, A., and Gonzalez-Rodriguez, J. 1991, Assignment of disulphide bonds in human platlet GPIIIa, *Biochem. J.* 274, 63-71.

Campanero, M.R., Pulido, R., Ursa, M.A., Rodriguez-Moya, M., deLandazuri, M.O., and Sanchez-Madrid, F. 1990, An alternative leukocyte homotypic adhesion mechanism, LFA-1/ICAM-1-independent, triggered through the human VLA-4 integrin, *J. Cell Biol.* 110, 2157-2165.

Carlos, T.M. and Harlan, J.M 1990, Membrane proteins involved in phagocyte adherence to endothelium, *Immunol. Rev.* 114, 5-28.

Carter, W.G., Kaur, P., Gil, S.G., Gahr, P.J., and Wayner, E.A. 1990a, Distinct functions for integrins $\alpha_3\beta_1$ in focal adhesions and $\alpha_6\beta_4$ /bullous pemphigoid antigen in a new stable anchoring contact (SAC) of keratinocytes: Relation to hemidesmosomes, *J. Cell Biol.* 111, 3141-3154.

Carter, W.G., Wayner, E.A., Bouchard, T.S., and Kaur, P. 1990b, The role of integrins alpha2 beta1 and alpha3 beta1 in cell-cell and cell-substrate adhesion of human epidermal cells, *J. Cell Biol.* 110, 1387-1404.

Chen, W-T., Chen, J-M., and Mueller, S.C. 1986, Coupled expression and colocalization of 140K cell adhesion molecules, fibronectin, and laminin during morphogenesis and cytodifferentiation of chick lung cells, *J. Cell Biol.* 103, 1073-1090.

Cohen, J., Nurcombe, V., Jeffrey, P., and Edgar, D. 1989, Developmental loss of functional laminin receptors on retinal ganglion cells is regulated by their target tissue, the optic tectum, *Develop.* 107, 381-387.

D'Souza, S.E., Ginsberg, M.H., Lam, S.C-T., and Plow, E. 1988, Chemical cross-linking of arginyl-glycyl-aspartic acid peptides on adhesion receptors on platelets, *Science* 242, 91-93.

D'Souza, S.E., Ginsberg, M.H., Matsueda, G.R., and Plow, E.F. 1991, A discrete sequence in a platelet integrin is involved in ligand recognition, *Nature* 350, 66-68.

Damsky, C.H., Knudsen, K.A., Bradley, D., Buck, C.A., and Horwitz, A.F. 1985, Distribution of the cell-substratum attachment (CSAT) antigen on myogenic and fibroblastic cells in culture, *J. Cell Biol.* 100, 1528-1539.

DeSimone, D.W. and Hynes, R.O. 1988, Xenopus laevis Integrins: Structural conservation and evolutionary divergence of integrin β subunits, *J. Biol. Chem.* 263, 5333-5340.

Elices, M.J., Osborn, L., Takada, Y., Course, C., Luhowskyl, S., Hemler, M., and Lobb, R. 1990, VCAM-1 on activated endothelium interacts with the leukocyte integrin VLA-4 at a site distinct from the VLA-4/fibronectin binding site, *J. Immunol.* 60, 577-584.

Elices, M.J., Urry, L.A., and Hemler, M.E. 1991, Receptor functions for the integrin VLA-3: Fibronectin, collagen, and laminin binding are differentially influenced by Arg-Gly-Asp peptide and by divalent cations, *J. Cell Biol.* 112, 169-181.

Feldman, L.E., Shin, K.C., Natale, R.B., and Todd, R.F.,III 1991, β_1 Integrin expression on human small cell lung cancer cells, *Cancer Res.* 51, 1065-1070.

Fingerman, E. and Hemler, M.E. 1988, Regulation of proteins in the VLA cell substrate adhesion family: Influence of cell growth conditions on VLA-1, VLA-2, and VLA-3 expression, *Exp. Cell Res.* 177, 132.

Frelinger, A.L., Cohen, I., Plow, E.F., Smith, M.A., Roberts, J., Lam, S.C.-T., and Ginsberg, M.H. 1990, Selective inhibition of integrin function by antibodies specific for ligand-occupied receptor conformers, *J. Biol. Chem.* 265, 6346-6352.

Ginsberg, M.H., Loftus, J.C., and Plow, E.F. 1988, Cytoadhesins, integrins, and platelets, *Thromb. Haemostasis* 59, 1-6.

Guan, J-L. and Hynes, R.O. 1990, Lymphoid cells recognize an alternatively spliced segment of fibronectin via the integrin receptor $\alpha_4\beta_1$, *Cell* 60, 53-61.

Hayashi, Y., Haimovich, B., Reszka, A., Boettiger, D., and Horwitz, A. 1990, Expression and function of chicken integrin β_1 subunit and its cytoplasmic domain mutants in mouse NIH 3T3 Cells, *J. Cell Biol.* 110, 175-184.

Hemler, M.E. 1990, VLA proteins in the integrin family: Structures, functions and their role in leukocytes, *Annu. Rev. Immunol.* 8, 365-400.

Hemler, M.E., Crouse, C., and Sonnenberg, A. 1989, Association of the VLA α_6 subunit with a novel protein, *J. Biol. Chem.* 264, 6529-6536.

Hogervorst, F., Kuikman, I., von dem Borne, A.E.G.R., and Sonnenberg, A. 1990, Cloning and sequence analysis of beta-4 cDNA: An integrin subunit that contains a unique 118 kd cytoplasmic domain, *EMBO J.* 9, 765-770.

Holers, V.M., Ruff, T.G., Parks, D.L., McDonald, J.A., Ballard, L.L., and Brown, E.J. 1989, Molecular cloning of a murine fibronectin receptor and its expression during inflammation: Expression of VLA-5 is increased in activated peritoneal macrophages in a manner discordant from major histocompatibility complex class II, *J. Exp. Med.* 169, 1589-1605.

Holzman, B. and Weissman, I.L. 1989, Integrin molecules involved inlymphocyte homing to Peyer's patches, *Immunol. Rev.* 108, 45-60.

Horvath, A.R., Elmore, M.A., and Kellie, S. 1990, Differential tyrosine-specific phosphorylation of integrin in Rous sarcoma virus transformed cells with differing transformed phenotypes, *Oncogene* 5, 1349-1357.

Hynes, R.O. 1987, Integrins, a family of cell surface receptors, *Cell* 48, 549-555.

Inghirami, G., Grignani, F., Sternas, L., Lombardi, L., Knowles, D.M., and Dalla-Favera, R. 1990, Down-regulation of LFA-1 adhesion receptors by C-*myc* oncogene in human B lymphoblastoid cells, *Science* 250, 682-686.

Isberg, R.R. and Leong, J.M. 1990, Multiple β_1 chain integrins are receptors for invasin, a protein that promotes bacterial penetration into mammalian cells, *Cell* 60, 861-871.

Koretz, K., Schlag, P., Boumsell, L., and Moller, P. 1991, Expression of VLA-α_2, VLA-α_6, and VLA-β_1 chains in normal mucosa and adenomas of the colon, and in colon carcinomas and their liver metastases, *Am. J. Pathol.* 138, 741-750.

Larjava, H., Peltonen, J., Akiyama, S.K., Yamada, S.S., Gralnick, H.R., Uitto, J., and Yamada, K.M. 1990, Novel function for β_1 integrins in keratinocyte cell-cell interactions, *J. Cell Biol.* 110, 803-815.

Larjava, H., Pelsonen, J., Akiyama, S., Yamada, S., Gralnick, J., Uitto, J., and Yamada, K. 1990 Novel function for beta 1 integrins in keratinocyte cell-cell interactions, *J. Cell Biol.* 110, 803-815.

Leininger, E., Roberts, M., Kenimer, J.G., Charles, I.G., Fairweather, N., Novotny, P., and Brennan, M.J. 1991, Pertactin, an Arg-Gly-Asp-containing *Bordetella pertussis* surface protein that promotes adherence of mammalian cells, *Proc. Natl. Acad. Sci.* 88, 345-349.

Leong, J.M, Fournier, R.S., and Isberg, R.R. 1990, Identification of the integrin binding domain of the *Yersinia pseudotuberculosis* invasin protein, *EMBO J.* 9, 1979-1989.

MacKrell, A., Blumberg, B., Haynes, S.R., and Fessler, J.H. 1988, The lethal myospheroid gene of drosophila encodes a membrane protein homologous to vertebrate integrin b subunits, *Proc. Natl. Acad. Sci.* 85, 2633-2637k.

Marcantonio, E.E., Guan, J-L., Trevithick, J.E., and Hynes, R.O. 1990, Mapping of the functional determinants of the integrin $\beta 1$ cytoplasmic domain by site-directed mutagenesis, *Cell Regulation* 1, 597-604.

Massague, J. 1990, The transforming growth factor-β family, *Annu. Rev. Cell Biol.* 6, 597-641.

McGregor, B., McGregor, J.L., Weiss, L.M., Wood, G.S., Hu, C-H., Boukerche, H., and Warnke, R.A. 1989, Presence of cytoadhesins (IIβ=IIIα-like glycoproteins) on human metastatic melanomas but not on benign melanocytes, *Am. J. Clin. Pathol.* 92, 495-499.

Menko, A.S. and Boettiger, D. 1987, Occupation of the extracelular matrix receptor, integrin, is a control point for myogenic differentiation *Cell* 51, 51-57.

Naidet, C., Semeriva, M., Yamada, K.M., and Thiery, J.P. 1987, Peptides containing the cell-attachment recognition signal Arg-Gly-Asp prevent gastrulation in *Drosophila* embryos, *Nature* 325, 348-350.

Nakamura, T., Takahashi, K., Fukazawa, T., Koyanagi, M., Yokoyama, A., Kato, H., Yagita, H., and Okumura, K. 1990, Relative contribution of CD2 and LFA-1 to murine T and natural killer cell functions, *J. Immunol.* 145, 3628-3634.

Neff, N.T., Lowrey, C., Decker, F C., Tovar, A., Damsky, C., Buck, C., and Horwitz, A.F. 1982, Monoclonal antibody detaches embryonic skeletal muscle from extracellular matrices, *J. Cell Biol.* 95, 654-666.

Nermut, M.V., Green, N.M., Eason, P., Yamada, S.S., and Yamada, K.M. 1988, Electron microscopy and structural model of human fibronectin receptor, *EMBO J.* 7, 4093-4099.

Neugebauer, K.M. and Reichardt, L.F. 1991, Cell-surface regulation of β_1-integrin activity on developing retinal neurons, *Nature* 350, 68-71.

Nicholson, L.J. and Watt, F.M. 1991, Decreased expression of fibronectin and the $\alpha_5\beta_1$ integrin during terminal differentiation of human keratinocytes, *J. Cell Sci.* 98, 225-232.

Otey, C., Pavalko, F., and Burridge, K. 1990, An interaction between α-actinin and the β_1 integrin subunit *in vitro*, *J. Cell Biol.* 111, 721-730.

Patel, V. and Lodish, H. 1985, The fibronectin receptor on mammalian erythroid precursor cells: Characterization and developmental regulation, *J. Cell Biol.* 102, 449-456.

Peltonen, J., Larjave, H., Jaakkola, S., Gralnick, H., Akiyama, S., Yamada, S., Yamada, K., and Uitto, J. 1989, Localization of integrin receptors for fibronectin, collagen and laminin in human skin, *J. Clin. Invest.* 84, 1916-1923.

Phillips, D.R., Charo, I.F., Parise, L.V., and Fitzgerald, L.A. 1988, The platelet membrane glycoprotein IIb-IIIa complex, *Blood* 71, 831-843.

Pierschbacher, M.D. and Ruoslahti, E. 1984, Cell attachment activity of fibronectin can be duplicated by small synthetic fragments of the molecule, *Nature* 309, 30-33.

Pignatelli, M. and Bodmer, W.F. 1990, Integrin cell adhesion molecules and colorectal cancer, *J. Pathol.* 162, 95-97.

Pytela, R., Pierschbacher, M.D., and Ruoslahti, E. 1985, Identification and isolation of a 140 kd cell surface glycoprotein with properties expected of a fibronectin receptor, *Cell* 40, 191-198.

Reichardt, L.F. and Tomaselli, K.J. 1991, Extracellular matrix molecules and their receptors: Functions in neural development, *Annu. Rev. Neurosci.* 14, 531-570.

Relman, D., Tuomanen, E., Falkow, S., Golenbock, D.T., Saukkonen, K., and Wright, S.D. 1990, Recognition of a bacterial adhesin by an integrin: macrophage CR3 ($\alpha_M\beta_2$, CD11b/CD18) binds filamentous hemagglutinin of *Bordetella pertussis*, *Cell* 61, 1375-1382.

Rohana-Talamas, P., Wright, S.D., Lennartz, M.R., and Russell, D.G. 1991, Lipophosphoglycan from *Leishmania mexicana* promastigotes binds tomembers of the CR3, p150,95 and LFA-1 family ofleukocyte integrins, *J. Immunol.* 144, 4817-4824.

Rosen, H. and Gordon, S. 1990, The role of the type 3 complement receptor in the induced recruitment of myelomonocytic cells to inflammatory sites in the mouse, *Am. J. Respir. Cell Mol. Biol.* 3, 3-10.

Ruoslahti, E. and Pierschbacher, M.D. 1987, New perspectives in cell adhesion: RGD and integrins, *Science* 238, 491-497.

Ruoslahti, E. 1991, Integrins, *J. Clin. Invest.* 87, 1-5.

Russell, D.G. and Wright, S.D. 1988, Complement receptor type 3 (CR3) binds to an Arg-Gly-Asp-containing region of the major surface glycoprotein, gp63, of *Leishmania* promastigotes, *J. Exp. Med.* 168, 279-292.

Shimizu, Y., Van Seventer, G., Horgan, K., and Shaw, S. 1990a, Costimulation of proliferative responses of restig CD4+ T cells by the interaction of VLA-4 and VLA-5 with fibronectin or VLA-6 with laminin, *J. Immunol.* 145, 59-67.

Shimizu, Y., Van Seventer, G., Horgan, K., and Shaw, S. 1990b, Regulated expression and binding of three VLA (β_1) integrin receptors on T cells, *Nature* 345, 250-253.

Simpson, P.J., Todd, R., III, Fantone, J.C., Mickelson, J.K., Griffin, J.D., and Lucchesi, B.R. 1988, Reduction of experimental canine myocardial reperfusion injury by a monoclonal antibody (anti-Mo1, anti-CD11b) that inhibits leukocyte adhesion, *J. Clin. Invest.* 81, 624-9.

Smith, J.W. and Cheresh, D.A. 1988, The Arg-Gly-Asp-binding domain of the vitronectin receptor, *J. Biol. Chem.* 263, 18726-18731.

Smith, J.W., Vestal, D.J., Irwin, S.V., Burke, T.A., and Cheresh, D.A. 1990, Purification and functional characterization of integrin $\alpha_v\beta_5$, *J. Biol. Chem.* 265, 11008.

Solowska, J., Guan, J.G., Marcandonio, E., Buck, C.A., and Hynes, R.O. 1989, Expression of normal and mutant avian integrin subunits in rodent cells, *J. Cell Biol.* 109, 853-861.

Sonnenberg, A., Hogervorst, F., Osterop, A., and Veltman, F. 1988b, Identification and characterization of a novel antigen complex on mouse mammary tumor cells using a monoclonal antibody against platelet glycoprotein Ic, *J. Biol. Chem.* 263, 14030-14038.

Sonnenberg, A. and Linders, C.J. 1990, The $\alpha_6\beta_1$ (VLA-6) and $\alpha_6\beta_4$ protein complexes: tissue distribution and biochemical properties, *J. Cell Sci.* 96, 207-217.

Sonnenberg, A., Modderman, P.W., and Hogervotst, F. 1988a, Laminin receptor on platelets is the integrin VLA-6, *Nature* 336, 487-489.

Springer, T.A. 1990, The sensation and regulation of interactions with the extracellular environment: The cell biology of lymphocyte adhesion receptors, *Annu. Rev. Cell Biol.* 6, 359-402.

Stepp, M.A., Spurr-Michaud, S., Tisdale, A., Elwell, J., and Gipson, I.K. 1990, $\alpha_6\beta_4$ integrin heterodimer is a component of hemidesmosomes, *Proc. Natl. Acad. Sci. USA* 87, 8970-8974.

Tapley, P., Horwitz, A., Buck, C., Duggan, K., and Rohrschneider, L. 1989, Integrins isolated from Rous sarcoma virus-transformed chicken embryo fibroblasts, *Oncogene* 4, 325-333.

Tuomanen, E.I., Saukkonen, K., Sande, S., Cioffe, C., and Wright, S.D. 1989, Reduction of inflammation, tissue damage, and mortality in bacterial meningitis in rabbits treated with monoclonal antibodies against adhesion-promoting receptors of leukocytes, *J. Exp. Med.* 170, 959-969.

Vedder, N.B., Winn, R.K., Rice, C.L., Chi, E.Y., Arfos, K.-E., and Harlan, J.M. 1988, A monoclonal antibody to the adherence-promoting leukocyte glycoprotein, CD18, reduces organ injury and improves survival from hemorrhagic shock and resuscitation in rabbits, *J. Clin. Invest.* 81, 939-944.

Wardlaw, A.J., Hibbs, M.L., Stacker, S.A., and Springer, T.A. 1990, Distinct mutations in two patients with leukocyte adhesion deficiency and their functional correlates, *J. Exp. Med.* 172, 335-345.

Wawryk, S.O., Novotny, J.R., Wicks, I.P., Wilkinson, D., Maher, D., Salvaris, E., Welch, K., Fecondo, J., and Boyd, A.W. 1989, The role of the LFA-1/ICAM-1 interaction in human leukocyte homing and adhesion, *Immunol. Rev.* 108, 135-161.

Wayner, E.A., Garcia-Pardo, A., Humphries, M.J., McDonald, J.A., and Carter, W.G. 1989, Identification and characterization of the T lymphocyte adhesion receptor for an alternative cell attachment domain in plasma fibronectin, *J. Cell Biol.* 109, 1321-1330.

Werb, Z., Tremble, P., Behrendtsen, O., Crowley, E., and Damsky, C. 1989, Signal transduction through the fibronectin receptor induces collagenase and stromelysin gene expression, *J. Cell Biol.* 109, 877-889.

Wright, S.D., Levin, S.M., Jong, M.C.T., Chad, Z., and Kabbash, L.G. 1989, CR3 (CD11b/CD18) expresses one binding site for Arg-Gly-Asp-containing peptides and a second site for bacterial lipopolysaccharide, *J. Exp. Med.* 169, 175-183.

Yamada, K.M. and Kennedy, D.W. 1984, Dualistic nature of adhesive protein function: Fibronectin and its biologically active peptide fragments can autoinhibit fibronectin function, *J. Cell Biol.* 99, 29-36.

Zutter, M.M., Mazoujian, G., and Santoro, S.A. 1990, Decreased expression of integrin adhesive protein receptors in adenocarcinoma of the breast, *Am. J. Pathol.* 137, 863-870.

Molecular Interactions between Human Rhinoviruses and the Adhesion Receptor ICAM-1

Richard J. Colonno

Introduction

Human rhinoviruses (HRVs), members of the *Picornaviridae*, are the major causative agents of one of the more elusive diseases known to man, namely the common cold (Gwaltney, 1982). To date, there are 102 recognized serotypes that have been isolated and shown to be antigenically distinct (Hamparian *et al.*, 1987). HRVs are non-enveloped viruses that contain four non-glycosylated structural proteins, designated VP1, VP2, VP3, and VP4, which form a protein capsid with icosahedral symmetry. Within the viral capsid lies a single-stranded genome RNA which serves as a monocistronic mRNA for the synthesis of the 4 structural and 7 non-structural proteins of the virus. Upon entry into a cell, the RNA genome is translated into a large polyprotein which is subsequently cleaved by two viral proteases encoded within the polyprotein (Palmenberg, 1987). The genome RNA of picornaviruses contains all the information needed to initiate a viral infection since transfection of cells with the genome RNA alone will result in the production of infectious progeny virus (Mizutani *et al.*, 1985).

Receptor Families

HRVs have been divided into 3 families dependent on their requirement of cellular receptors as determined by competition binding and cell protection studies. Ninety-one of the 102 serotypes utilize a single cellular receptor and are designated the major group, while 10 (HRV-1A, -1B, -2, -29, -30, -31, -44, -47, -49, and -62) of the remaining 11 serotypes compete for a second receptor and comprise the minor group (Abraham *et al.*, 1984; Colonno *et al.*, 1986; Uncapher *et al.*, 1991). One serotype, HRV-87, appears to utilize neither the major nor minor group receptor and may represent a third receptor group. HRV-87, which is clearly not a major group virus, remains an anomaly since it shows cell binding tropisms similar to minor group viruses yet is unlike minor group viruses in requiring sialic acid for attachment (Uncapher *et al.*, 1991).

Characterization and Cloning of the Major Group Receptor

Characterization of the HRV major group receptor was made possible by the isolation of a murine monoclonal antibody, designated MAb 1A6, which recognized the major group receptor and inhibited virus binding (Colonno *et al.*, 1986). The major group viruses exhibit an absolute requirement for this receptor since blocking attachment with MAb 1A6 was cytoprotective despite high titer viral challenge (Colonno *et al.*, 1986). *In vivo* chimpanzee and human clinical trials involving MAb 1A6 have also demonstrated that this receptor is involved in HRV infection of the nasal cavity (Colonno *et al.*, 1987; Hayden *et al.*, 1988).

Using MAb 1A6 immunoaffinity chromatography, a 90 kDa surface glycoprotein was isolated from HeLa cells which had a pI of 4.2 (Tomassini *et al.*, 1986; Tomassini *et al.*, 1989b). Carbohydrates accounted for 30% of the molecular mass of the protein and 7 N-linked glycosylation sites were predicted based on partial digestion with N-glycanase (Tomassini *et al.*, 1989b). Digestion with neuraminidase caused a downward mobility shift of 10 kDa on SDS polyacrylamide gels, revealing the presence of sialic acid in the oligosaccharide component of the receptor protein (Tomassini *et al.*, 1989b). Lectin binding studies showed that wheat germ lectin, which has a specificity for sialic acid, inhibited attachment of major group viruses to HeLa cell membranes, thus confirming that sialic acid was a component of the HRV cellular receptor (Tomassini *et al.*, 1989b). Further digestion of desialyated receptor with beta-galactosidase resulted in an additional downward mobility shift of 2 kDa on SDS polyacrylamide gels, indicating the successive linkage of beta-galactose to sialic acid in the oligosaccharide chain.

Chemical and enzymatic digestions of the isolated receptor protein yielded 6 peptides that gave discernible sequence ranging from 5 to 23 amino acid residues and cumulatively representing 84 residues (Tomassini *et al.*, 1989a). Using degenerate oligonucleotides, representing deduced peptide sequence as probes, four overlapping clones were identified and joined together to generate a single clone (pHRVr1) containing a 3 kb insert (Tomassini *et al.*, 1989a). The full-length clone had a single large open reading frame initiating at nucleotide 72 that encoded 532 amino acids, and contained a 1333 nucleotide 3' noncoding region. *In vitro* translation of RNA transcribed from pHRVr1 resulted in a single polypeptide of 55 kDa that is in close agreement to the 54-60 kDa size found for deglycosylated receptor protein (Tomassini *et al.*, 1989a; Tomassini *et al.*, 1989b).

DNA sequencing of pHRVr1 revealed a striking homology to the intercellular adhesion molecule-1 (ICAM-1) (Staunton *et al.*, 1989; Tomassini *et al.*, 1989a). The identity of ICAM-1 as the major HRV group receptor was first recognized by Greve *et al.* (1989). In addition to the virtually identical sequence homology, both the major HRV receptor and ICAM-1 ligand proteins have equivalent mass, tissue distribution, and carbohydrate moieties (Colonno *et al.*, 1986; Dustin *et al.*, 1988; Staunton *et al.*, 1988; Tomassini *et al.*, 1989b).

To confirm that the ICAM-1 receptor was the functional receptor for the major group of HRVs, stable Vero cells were generated by transfection of an ICAM-1 plasmid containing an upstream SV40 early promoter (pSVL-HRVr1) followed by neomycin selection (Colonno *et al.*, 1990). Radiolabeled HRV-36, a major group virus, showed 24.9% binding to the selected Vero-ICAM+ cells in contrast to only the 3.2% binding observed when untransfected Vero cells were used (Colonno *et al.*, 1990). The binding to Vero-ICAM+ cells is specific since addition of MAb 1A6 reduced binding of HRV-36 to 2.6%. To determine if major group viruses could infect the Vero-ICAM+ cells, 10 major group serotypes were used to infect Vero and Vero-ICAM+ cells. While none of the viruses tested were able to infect ICAM- Vero cells, 9 of the 10 serotypes grew to varying degrees in the Vero ICAM+ cells (Colonno *et al.*, 1990).

ICAM-1

ICAM-1 is a cell surface ligand for lymphocyte function-associated antigen 1 (LFA-1), and the interaction between these molecules plays an important role in several immunological and inflammatory functions mediated by leukocyte adhesion (Dustin *et al.*, 1988; Makgoba *et al.*, 1988). The ICAM-1 ligand is a member of the immunoglobulin supergene family and is predicted to have 5 homologous immunoglobulin-like domains defined by amino acids 1-88, 89-185, 186-284, 285-385, and 386-453 (Staunton *et al.*, 1988). In addition, ICAM-1 is closely related to 2 adhesion proteins of the adult nervous system, namely neural cell adhesion molecule and myelin-associated glycoprotein (Simmons *et al.*, 1988; Staunton *et al.*, 1988).

The significance of HRVs utilizing receptors having immunological and inflammatory functions is unclear. It has been reported that HRV infection involves a very limited number of cells in the nasal epithelium, and that the clinical symptomatology associated with a cold may instead result from the generation of inflammatory mediators such as kinins (Naclerio *et al.*, 1988). It is tempting to speculate that HRV interaction with ICAM-1 somehow plays an important role in the production of kinins and/or other mediators detected during common cold infections.

Mapping the Binding Domains of ICAM-1

The prediction that ICAM-1 is structurally related to members of the immunoglobulin supergene family suggests that ICAM-1 is closely related to the CD4 receptor and to the immunoglobulin-like protein recently identified as the cellular receptor for poliovirus (Mendelsohn *et al.*, 1989). The CD4 receptor utilized by the human immunodeficiency virus (HIV) (Jameson *et al.*, 1988) is postulated to have 4 domains while the poliovirus receptor is estimated to have 3 homologous immunoglobulin-like domains (Mendelsohn *et al.*, 1989). Several laboratories have mapped the cellular binding site for HIV gp120 to a short region of the CD4 receptor that resides within the first 53 amino acids of the N-terminal domain (Jameson *et al.*, 1988; Peterson *et al.*, 1988). Anti-CD4 MAbs (OKT4A, Leu3A) that block HIV attachment also map to this same region (Jameson *et al.*, 1988; Landau *et al.*, 1988; Peterson *et al.*, 1988) and suggest that MAbs capable of abrogating virus attachment can be utilized to map the site of virus interaction.

Studies were therefore undertaken to map the ICAM-1 binding sites of 3 MAbs, including MAb 1A6, that efficiently block attachment of the major group of HRVs and are cytoprotective. Further analysis of these anti-receptor MAbs indicates that they bind to non-overlapping sites on ICAM-1 and are equally effective at blocking virus attachment (Lineberger *et al.*, 1990). Our approach differed from previous studies in that we employed an in vitro transcription/translation system to generate fragments of the ICAM-1 receptor. The system was further optimized by the inclusion of canine microsomal membranes and glutathione to enhance the generation of functional molecules. The microsomal membranes, which are only capable of core glycosylation, were found to be required for proper folding and/or glycosylation since immunoprecipitation experiments showed that only the proteins that had transversed the membranes were able to be immunoprecipitated by the 3 MAbs (Lineberger *et al.*, 1990). Interestingly, this was true for the shortest ICAM-1 fragment (Domain 1) which is not predicted to have glycosylation sites, and further suggests that translation in the presence of microsomal membranes has a conformational effect on in vitro synthesized ICAM-1.

Experiments utilizing the subset of ICAM-1 polypeptides clearly showed that all 3 of the anti-ICAM-1 MAbs assayed were able to recognize the smallest fragment tested (Lineberger *et al.*, 1990). This fragment represented only the first 82 amino acids of the mature ICAM-1 molecule and encompassed only the first of 5 predicted domains. It is interesting to note that the 82 amino acid long fragment contains 2 putative SH bonds bridging Cys residues at positions 21 and 25 with Cys residues 65 and 69, respectively (Tomassini *et al.*, 1989a). These SH bridges appear to be important in determining the conformation of domain 1, since shorter fragments deleting Cys residues 65 and 69 result in fragments no longer recognized by any of the MAbs (Lineberger *et al.*, 1990). Attempts to show specific virus binding with either the full-length protein or the domain 1 fragment translated *in vitro* were unsuccessful (Lineberger *et al.*, 1990).

Virion Attachment Site of Major Group Viruses

Rossmann *et al.* (1985) previously suggested that a deep crevice found on the surface of the virus may be involved in receptor interaction. It was reasoned that such a location for this highly-conserved site would exclude the production and/or accessibility of cross-neutralizing immunoglobulins which might be generated by the infected host. In addition to the concept of antibody exclusion, there are several lines of circumstantial evidence which are supportive. First, is the finding that 91 HRV serotypes utilize a highly-conserved attachment site to which the host immune system fails to generate antibodies capable of cross neutralizing a vast number of serotypes (Uncapher *et al.*, 1991). Second, is the failure to generate anti-idiotypic antibody to anti-receptor MAbs which are capable of recognizing HRVs (Colonno, 1987). Lastly, is the failure to generate viable mutants of HRVs which are capable of bypassing the MAb 1A6 blockade (Colonno *et al.*, 1986). Mutations within the core structure, unlike the surface projections forming the virion neutralization sites (Sherry *et al.*, 1986), are more likely to result in capsid instability and non-infectious virus.

The most direct evidence for the "canyon hypothesis" comes from recombinant DNA studies in which a series of single amino acid changes were introduced into the VP1 protein of HRV-14 at 4 amino acid residues that map to the deepest part of the canyon (Colonno *et al.*, 1988). Replacing the Lys at position 103 with another positively-charged amino acid, Arg, or with an uncharged Ile, had no significant effect on virus binding. However, if a neutral, polar Asn residue was inserted at position 103, then the binding of the mutant was negatively affected. The changes at position 155 were very interesting. Replacing the Pro with a smaller amino acid, Gly, resulted in a nine-fold increase in binding affinity while a change to a larger residue, Tyr, apparently is lethal. Positions 220 and 223 were the most sensitive to substitution. Substituting the weakly-charged His at position 220 with nonpolar residues, such as Ile or Trp, generated mutant viruses with drastically reduced binding affinities. Similarly, replacement of the hydroxyl-containing Ser at position 223 with an Ala or Thr resulted in mutants with decreased binding affinities. It is clear that the proper blend of size, charge, and hydrogen bonding properties is critical for receptor protein interaction with this region of the canyon.

The binding phenotypes displayed by the HRV-14 mutants represent conclusive evidence that the canyon floor is involved in receptor interaction. Whether the observed binding affinities were the result of simple charge changes or reflect gross anatomical changes in the floor of the canyon is not known. However, the possibility that these changes could have affected structures outside of the canyon is remote since a rippling effect, involving numerous amino acids within the wall of the canyon, would be needed to significantly alter external structures some 22 angstroms away. The region selected for amino acid substitutions is one that is highly conserved not only among HRV serotypes but also in a number of other picornaviruses such as poliovirus and coxsackieviruses A and B. It is highly unlikely that this particular region is involved in determining receptor specificity. Instead, it is more probable that this conserved region interacts with a common domain found on several receptors and that it is the shape and contour of the canyon entrance which defines the accessibility of a receptor protein into the viral canyon.

Additional support for the canyon model has come from studies using a series of compounds that interfere with virus uncoating. These small hydrophobic molecules were shown to enter a small pore located at the bottom of the canyon. In the case of the major group serotype, HRV-14, these drugs cause amino acids located at the floor of the canyon, including the four amino acid residues mentioned above, to move upward into the canyon and interfere with virus attachment (Prevear *et al.*, 1989).

Interactions between ICAM-1 and HRVs

Computer modeling of ICAM-1 domain 1 proposes that the N-terminal half of domain 1 could occupy the HRV canyon and interact with amino acids located at the bottom of the canyon (Giranda *et al.*, 1990). Although highly speculative in nature, this docking of ICAM-1 and its predicted interactions with the virion canyon are supported by the mutagenesis studies above and those performed on ICAM-1 (Staunton *et al.*, 1990).

It has also been postulated that the cellular receptor site, capable of interacting with the canyon, may be a multimeric complex composed of five monomeric, 90 kDa receptor proteins. This concept is supported by the finding that solubilized HRV receptor exists as a 440 kDa complex in HeLa cells (Tomassini *et al.*, 1986). However, direct evidence for preferential binding of multimers vs. monomers in virus binding has yet to be achieved. HRV attachment appears to be a reversible process since virus released from cellular receptors by MAb 1A6 or other viruses is completely intact and infectious (Abraham *et al.*, 1988). This result, in contrast to recent studies with poliovirus (Kaplan *et al.*, 1990), indicates that the initial interaction of the viral canyon with the cellular receptor protein does not alter the conformation of the viral capsid. In addition, it is now apparent that the myristic acid component of VP4, recently described by Chow *et al.* (1987), is not involved in the interaction between virus and cellular receptors (Abraham *et al.*, 1988).

Cellular Receptors for Other HRVs

The finding that the canyon is highly conserved among a number of picornaviruses belonging to different receptor families led to the supposition that the receptor molecules that interact with these different viruses have a structural binding domain in common (Colonno *et al.*, 1988).

It was further suggested that the specificity of the receptor protein utilized for attachment was determined by the rim of the canyon, since this region of the capsid was highly variable and thus would define the structural requirements of receptor interaction with the virus.

MAbs that block the attachment of HRV-87 and the minor group viruses have yet to be isolated. While definitive experiments have yet to be performed, a putative 120 kDa minor group receptor protein has been identified using a virus blot technique in which radiolabeled virus is used to detect detergent-solubilized membrane proteins following SDS-PAGE (Mischak *et al.*, 1988). Lectin competition studies have demonstrated that wheat germ lectin inhibited the attachment of both major and minor group viruses (Tomassini *et al.*, 1989b) and along with the results obtained for HRV-87 (Uncapher *et al.*, 1991), suggest that sialic acid may be a component of all three of the cellular receptors for these viruses. Sialic acid does not appear to play a functional role in the attachment of major and minor group viruses since neuraminidase treatment of HeLa cell monolayers fails to block virus binding and infection (Uncapher *et al.*, 1991). The opposite was true for HRV-87 which required sialic acid for attachment and infection (Uncapher *et al.*, 1991). Based on the results above and the knowledge that a similar virion canyon also exists in the minor group virus, HRV-1A (Kim *et al.*, 1989), it seems logical to expect that the minor group receptor will also be a member of immunoglobulin supergene family. While the RNA genome of HRV-87 remains to be sequenced, it appears likely that the virion attachment site and receptor for this virus may have unique structural characteristics.

References

Abraham, G., and Colonno, R.J., 1983, Many rhinovirus serotypes share the same cellular receptor, *J. Virol.* 51:340-345.

Abraham, G., and Colonno, R.J., 1988, Characterization of human rhinoviruses displaced by an anti-receptor monoclonal antibody, *J. Virol.* 62:2300-2306.

Chow, M., *et al.*, 1987, Myristylation of picornavirus capsid protein VP4 and its structural significance, *Nature (London)* 327:482-486.

Colonno, R.J., 1987, Cell surface receptors for picornaviruses, *BioEssays* 5:270-274.

Colonno, R.J., Callahan, P.L., and Long, W.J., 1986, Isolation of a monoclonal antibody that blocks attachment of the major group of human rhinoviruses, *J. Virol.* 57:7-12.

Colonno, R.J., Tomassini, J.E., and Callahan, P.L., 1987, Isolation and characterization of a monoclonal antibody which blocks attachment of human rhinovirus, In: Positive Strand RNA Viruses (Eds. M.A. Brinton and R. Rueckert) Vol 54:93-102.

Colonno, R.J., *et al.*, 1988, Evidence for the direct involvement of the rhinovirus canyon in receptor binding, *Proc. Natl. Acad. Sci. USA* 85:5449-5453.

Colonno, R.J., *et al.*, 1990, The major-group rhinoviruses utilize the intercellular adhesion molecule 1 ligand as a cellular receptor during infection, In: New Aspects of Positive-Strand RNA Viruses (Eds. M.A. Brinton & F. X. Heinz) 257-261.

Dustin, M.L., Staunton, D.W., and Springer, T.A., 1988, Supergene families meet in the immune system, *Immunol. Today* 9:213-215.

Giranda, V.L., Chapman, M.S., and Rossmann, M.G., 1990, Modeling of the human intercellular adhesion molecule-1, the human rhinovirus major group receptor, In: Proteins: Structure, Function, and Genetics 7:227-233.

Greve, J.M., *et al.*, 1989, The major human rhinovirus receptor is ICAM-1, *Cell* 56:839-847.

Gwaltney, J.M., Jr., 1982, Rhinoviruses, In: Viral Infection of Man: Epidemiology and Control (ed. E.A. Evans) 491-517.

Hamparian, V.V., *et al.*, 1987, A collaborative report: rhinoviruses--extension of the numbering system from 89 to 100, *Virol.* 159:191-192.

Hayden, F.G., Gwaltney, Jr., J.M., and Colonno, R.J., 1988, Modification of experimental rhinovirus colds by receptor blockade, *Antiviral Res.* 9:233-247.

Jameson, B.A., *et al.*, 1988, Location and chemical synthesis of a binding site for HIV-1 on the CD4 protein, *Science* 240:1335-1339.

Kaplan, G., Freistadt, M.S., and Racaniello, V.R., 1990, Neutralization of poliovirus by cell receptors expressed in insect cells, *J. Virol.* 64:4697-4702.

Kim, S., *et al.*, 1989, Crystal structure of human rhinovirus serotype 1A (HRV1A), *J. Mol. Biol.* 210:91-111.

Landau, N.R., Warton, M., and Littman, D.R., 1988, The envelope glycoprotein of the human immunodeficiency virus binds to the immunoglobulin-like domain of CD4, *Nature* 334:159-162.

Lineberger, D.W., *et al.*, 1990, Antibodies that block rhinovirus attachment map to domain 1 of the major group receptor, *J. Virol.* 64: 2582-2587.

Makgoba, M.W., *et al.*, 1988, Functional evidence that intercellular adhesion molecule-1 (ICAM-1) is a ligand for LFA-1 dependent adhesion in T cell-mediated cytotoxicity, *Eur. J. Immunol.* 18:637-640.

Mendelsohn, C.L., Wimmer, E., and Racaniello, V., 1989, Cellular receptor for poliovirus: molecular cloning, nucleotide sequence, and expression of a new member of the immunoglobulin superfamily, *Cell* 56:855-865.

Mischak, H.C., *et al.*, 1988, Detection of the human rhinovirus minor group receptor on renaturing Western blots, *J. Gen. Virol.* 69:2653-2656.

Mizutani, S., and Colonno, R.J., 1985, *In vitro* synthesis of an infectious RNA from cDNA clones of human rhinovirus type 14, *J. Virol.* 56:628-632.

Naclerio, R.M., *et al.*, 1988, Kinins are generated during experimental rhinovirus colds, *J. Infect. Dis.* 157:133-142.

Palmenberg, A.C., 1987, Picornaviral processing: some new ideas, *J. Cell. Biochem.* 33:191-198.

Peterson, A., and Seed, B. 1988, Genetic analysis of monoclonal antibody and HIV binding sites on the human lymphocyte antigen CD4, *Cell* 54:65-72.

Prevear, D.C., *et al.*, 1989, Conformational change in the floor of the human rhinovirus canyon blocks adsorption to HeLa cell receptors, *J. Virol.* 63:2002-2007.

Rossmann, M.G., *et al.*, 1985, Structure of a human common cold virus and functional relationship to other picornaviruses, *Nature* 317:145-153.

Sherry, B., *et al.*, 1986, Use of monoclonal antibodies to identify four neutralization immunogens on a common cold picornavirus, human rhinovirus 14, *J. Virol.* 57:246-257

Simmons, D., Makgoba, M.W., and Seed, B., 1988, ICAM, an adhesion ligand of LFA-1, is homologous to the neural cell adhesion molecule NCAM, *Nature* 331:624-627.

Staunton, D.E., *et al.*, 1988, Primary structure of ICAM-1 demonstrates interaction between members of the immunoglobulin and integrin supergene families, *Cell* 52:925-933.

Staunton, D.E., *et al.*, 1989, A cell adhesion molecule, ICAM-1, is the major surface receptor for rhinoviruses, *Cell* 56:849-853.

Staunton, D.E., *et al.*, 1990, The arrangement of the immunoglobulin-like domains of ICAM-1 and the binding sites for LFA-1 and rhinovirus, *Cell*, 61:243-254.

Tomassini, J.E., and Colonno, R.J., 1986, Isolation of a receptor protein involved in attachment of human rhinoviruses, *J. Virol.* 58:290-295.

Tomassini, J.E., Maxson, T.R., and Colonno, R.J., 1989b, Biochemical characterization of a glycoprotein required for rhinovirus attachment, *J. Biol. Chem.* 264:1656-1662.

Tomassini, J.E., *et al.*, 1989a, cDNA cloning reveals that the major group rhinovirus receptor on HeLa cells is intercellular adhesion molecule 1, *Proc. Natl. Acad. Sci. USA* 86:4907-4911.

Uncapher, C.R., DeWitt, C.M., and Colonno, R.J., 1991, The major and minor group receptor families contain all but one human rhinovirus serotype, *Virol.* in press.

Cell Surface Receptors Required for Herpes simplex Virus Infection Include Heparan Sulfate Glycosaminoglycans

Patricia G. Spear, Mei-Tsu Shieh, Betsy C. Herold, and Darrell WuDunn

Introduction

Herpes simplex virus (HSV) is one of the several biologically distinct herpes viruses that cause disease in humans. The most common manifestations of HSV disease include keratitis and cutaneous lesions, such as cold sores and fever blisters, and similar lesions on genital organs. Rarely, the virus can induce more serious disease such as encephalitis and disseminated infections affecting several organ systems. HSV is not eliminated following the healing of lesions but persists as latent virus in the neurons of sensory or autonomic ganglia connecting with sites of viral inoculation or replication. Reactivation of latent virus is responsible for the recurrences of lesions that afflict many people. There are two serotypes of HSV, designated HSV-1 and HSV-2. HSV-1 is most commonly associated with facial lesions, keratitis and encephalitis in adults whereas HSV-2 is most commonly associated with genital lesions and neonatal infections (Corey *et al.*, 1988).

HSV has a DNA genome of about 150 kbp. In the virion, this DNA is enclosed within an icosahedral capsid, which is in turn enclosed within a membranous envelope composed of lipids and viral glycoproteins (Roizman *et al.*, 1990). There are at least seven, and probably more, distinct viral glycoproteins present in the virion envelope. Some of these glycoproteins play key roles in the adsorption of virus to cells and in penetration of virus into cells, as will be discussed below. The others may have only peripheral roles in infectivity.

Overview of the Requirements for HSV Infectivity

Infectivity may be defined as the ability of a virus to invade a cell so as to subvert the cell to the expression of at least some viral genes. For all enveloped animal viruses studied to date, infectivity requires the adsorption of virus to specific cell surface receptors followed by fusion of the viral envelope with a cell membrane. This fusion may occur with the plasma membrane of the cell or may occur with the membrane of an endocytic vesicle after endocytosis of the virion. In the case of HSV, infectious entry of virus can occur by fusion of the virion envelope with the plasma membrane (Fuller *et al.*, 1987; Wittels *et al.*, 1991).

At least four of the several glycoproteins found in the envelope of HSV participate in the entry of HSV into cells (Fig. 1). The glycoprotein that is principally responsible for the adsorption of virus to cells is gC (Herold *et al.*, 1991). There appears to be a back-up mechanism for virion adsorption to cells. Viral mutants devoid of gC have detectable infectivity, although their infectivity is significantly impaired due in large part to impairment in adsorption. The possibility exists that gB mediates the adsorption of gC-negative mutants to cells, as will be discussed below. Whatever the role of gB in virus adsorption, this glycoprotein also has a role in viral penetration. Three of the HSV glycoproteins (gB, gD and gH) are essential for events occurring subsequent to adsorption and prior to viral penetration. This conclusion is based on studies done with viral mutants and with monoclonal antibodies capable of neutralizing HSV infectivity by blocking viral penetration (Sarmiento *et al.*, 1979; Cai *et al.*, 1988; Ligas *et al.*, 1988; Desai *et al.*, 1988; Fuller *et al.*, 1987; Highlander *et al.*, 1987; Highlander *et al.*, 1988; Fuller *et al.*, 1989).

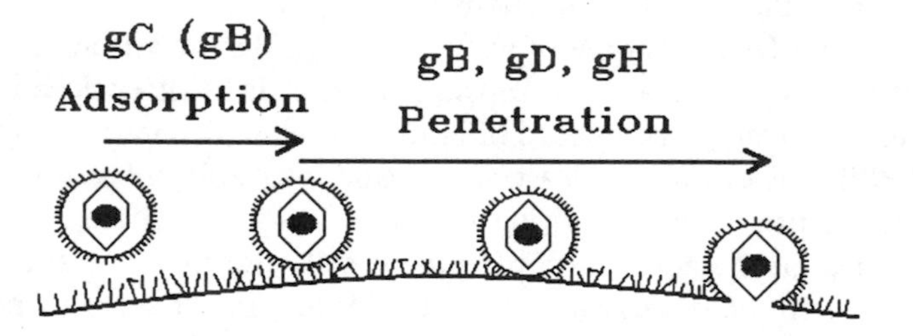

Fig. 1. The HSV glycoproteins that mediate adsorption of virus to cells and viral penetration. Both gC and gB are heparin-binding glycoproteins. In the case of wild-type virions, gC mediates the attachment of virions to cell surface heparan sulfate. In gC-negative virions, which have significantly reduced infectivity, gB may mediate adsorption, also to cell surface heparan sulfate. The three glycoproteins gB, gD and gH have essential, but as yet undefined roles, in events occurring between adsorption and fusion of the virion envelope with the cell membrane.

Attention will be focused here on describing the results that led to the identification of cell surface receptors for HSV and of the virion glycoproteins that mediate the attachment of virus to these receptors.

Heparan Sulfate Glycosaminoglycans Are Receptors for HSV

Proteoglycans are integral components of the plasma membranes of most eukaryotic animal cells (Hook *et al.*, 1984), with the possible exception of some non-sessile cells of the hematopoietic system. The carbohydrate moieties of cell surface proteoglycans include glycosaminoglycans (GAGs), the principal forms of which are heparan sulfate, chondroitin sulfate and dermatan sulfate (Fig. 2). Heparin is structurally very similar to heparan sulfate except that heparin is made by specialized cells for secretion and is usually more highly sulfated than heparan sulfate.

The early observations that heparin could inhibit HSV infection (Nahmias *et al.*, 1964; Takemoto *et al.*, 1964; Vaheri, 1964) led to studies showing that both HSV-1 and HSV-2 make their initial contact with cells by binding to cell surface heparan sulfate and that soluble heparin can inhibit this binding. Some of the published results (WuDunn *et al.*, 1989) in support of these conclusions may be summarized as follows: (i) Heparin inhibited HSV infection by blocking the adsorption of virus to cells. (ii) The binding of virions to heparin-Sepharose and blocking of this binding by soluble heparin could be demonstrated. (iii) The binding of HSV to cells and infection of the cells was significantly inhibited by pretreatment of the cells with heparitinase or heparinase, both of which can hydrolyze heparan sulfate. Treatment of cells with chondroitin lyase, on the other hand, had no effect on ability of the cells to bind virus and to be infected. Moreover, cells treated with any of these enzymes retained full susceptibility to an unrelated virus such as Sendai virus.

Cell mutants defective in glycosaminoglycan synthesis have also been studied, in order to test further the hypothesis that the binding of HSV to cells requires the presence of cell surface heparan sulfate. Esko and collaborators (1985) isolated glycosaminoglycan mutants of Chinese hamster ovary (CHO) cells by screening for clones that failed to incorporate radioactive inorganic sulfate into macromolecules. Some of these mutants and their biochemical defects are described in Table 1.

In studies to be published elsewhere (M.-T. Shieh, D. WuDunn, R.I. Montgomery, J.D. Esko and P.G. Spear, manuscript submitted), it has been shown that radiolabeled HSV-1 or HSV-2 binds efficiently to wild-type CHO cells but not to any of the three mutants (CHO-745, CHO-761, CHO-677) that fail to produce heparan sulfate, despite the fact that one of these mutants (CHO-677) produces chondroitin sulfate. In addition, binding of HSV to the mutant that produces undersulfated heparan sulfate (CHO-606) was detectable but was significantly impaired. These results show that the binding of HSV to cells depends upon the presence of heparan sulfate, but

Xxx
Xxx
Xxx
Ser - Xylose - Galactose - Galactose - Glucuronic acid - $(A - B)_n$
Gly
Xxx
Xxx

	A	B
Heparan sulfate	Glucosamine (N-SO3- or N-Ac; O-sulfate)	Hexuronic acid (O-sulfate)
Chondroitin sulfate	Galactosamine (N-Ac; O-sulfate)	Glucuronic acid
Dermatan sulfate	Galactosamine (N-Ac; O-sulfate)	Iduronic acid (O-sulfate)

Fig. 2. Glycosaminoglycans (GAGs) found on cell surface proteoglycans (Hook *et al.*, 1984). The top part of this figure shows how the GAGs are linked to Ser residues in a core protein. The bottom part shows the composition of the repeating disaccharide unit found in each of the three major GAGs present on cell surface proteoglycans. In parentheses are noted the modifications of the amino sugar or hexuronic acid that can be detected in a variable fraction of the disaccharide units that make up each polymer.

not chondroitin sulfate, and that the degree of sulfation of the heparan sulfate chains influences the number or affinity of HSV receptors. Although CHO cells are not permissive for HSV replication, they do support an abortive infection with this virus. Specifically, expression of an immediate

Table 1. CHO cell lines obtained from Jeffrey D. Esko

			GAGs produced		
Mutant group	Strain	Biochemical deficiency	Heparan sulfate	Chondroitin sulfate	Reference
Wild-type	K1	None	Yes	Yes	
pgsA	745	Xylosyltransferase	No	No	Esko *et al.* (1985)
pgsB	761	Galactosyltransferase	No	No	Esko *et al.* (1987)
pgsD	677	N-acetyl-glucosaminyl- and glucuronosyltransferases	No	Yes	Esko *et al.*(1988)
pgsE	606	N-sulfotransferase	Yes[a]	Yes	Bame & Esko (1989)

[a]The GAGs produced are under-sulfated.

early viral gene, designated ICP4, can be detected (by immunofluorescence, for example). We have found, for both HSV-1 and HSV-2, that the amount of virus bound to the mutant or wild-type CHO cells listed in Table 1 correlates directly with the percentage of cells that can be induced to express ICP4.

Are Other Cell Surface Receptors Required for HSV Infection?

It seems likely that the binding of HSV to cell surface heparan sulfate triggers other reactions at the cell surface that lead to viral penetration. It is difficult to envision how the simple binding of virus to heparan sulfate could be sufficient to induce membrane fusion and viral penetration. The possibility exists that interactions between multiple HSV glycoproteins and multiple cell surface receptors may be required during the events leading up to membrane fusion and viral penetration into the cell. This is suggested by findings that (i) at least four glycoproteins participate in adsorption and penetration; (ii) at least three of these four glycoproteins form three morphologically distinct spikes in the virion envelope (Stannard *et al.*, 1987) (rather than forming a single hetero-oligomeric functional unit); and (iii) all events preparatory to membrane fusion can occur at the cell surface (Fuller *et al.*, 1987; Wittels *et al.*, 1991).

It has been proposed that, subsequent to the adsorption of HSV to a cell, gD must interact with a specific cell surface receptor in order for penetration to occur (Johnson *et al.*, 1988; Johnson *et al.*, 1990; Campadelli-Fiume *et al.*, 1988; Johnson *et al.*, 1989). This proposal was based on three observations. First, it was found that exposure of cells to UV-inactivated wild-type virions, but not to mutant virions devoid of gD, could block the entry of infectious virus into cells. Adsorption of the infectious virus was not inhibited, implying that the competition was for receptors required for penetration, not for adsorption (Johnson *et al.*, 1988). Second, it was shown that a secreted form of gD could bind to cells, albeit with low affinity (Johnson *et al.*, 1990). Third, it was found that cells transformed to express gD were resistant to HSV infection. The fact that binding of virus to the gD-expressing cells was not impaired suggested that there was competition between cell-associated gD and virion-associated gD for a cell receptor required for viral penetration (Campadelli-Fiume *et al.*, 1988; Johnson *et al.*, 1989).

A more recent study casts doubt on this interpretation of the gD-mediated interference phenomenon, however. It was shown that an HSV mutant resistant to the interference could be isolated; that the relevant mutation mapped to the gD gene; and that the mutated gD could not itself mediate interference (Campadelli-Fiume *et al.*, 1990). The authors' interpretation of these results was that a homotypic interaction between wild-type forms of cell-associated gD and virion-associated gD was responsible for the interference and that the "interference" domain of gD could be eliminated by mutation without lethal effects on essential function(s) of gD. Thus, it remains to be determined whether gD interacts with a specific cell surface receptor and, if so, the receptor has yet to be identified.

It has also been suggested that a receptor for fibroblast growth factor (FGF) might play a role in the entry of HSV into cells (Kaner *et al.*, 1990; Baird *et al.*, 1990). The preliminary observation which led to further studies was a finding that FGF inhibited HSV infection. Because members of the FGF family of proteins are heparin-binding proteins and also bind to high affinity protein receptors on cell surfaces (Burgess *et al.*, 1989), at least three interpretations of the anti-HSV activity seem reasonable. FGF could block the adsorption of HSV to heparan sulfate, as has been shown for another heparin-binding protein, platelet factor 4 (WuDunn *et al.*, 1989). Alternatively, FGF could block the adsorption of virus to the protein receptor for FGF or to both the low affinity (heparan sulfate) and high affinity (protein) FGF receptors.

The authors favored the latter interpretations, based on findings that CHO cells transfected to express a mouse FGF receptor bound more radiolabel from preparations of HSV than did control CHO cells (Kaner *et al.*, 1990). They also reported that anti-FGF antibodies could inhibit this binding of label (Baird *et al.*, 1990). It was concluded that HSV entered cells via the FGF receptor by "piggy-backing" on FGF itself. Crucial issues not addressed in this work include whether presence or absence of high affinity FGF receptors renders cells more or less susceptible to HSV infection and whether a physical association between HSV and FGF can be demonstrated.

We conclude that a role for high affinity FGF receptors in HSV adsorption or penetration has not been established. It remains to be determined whether cell surface receptors other than heparan sulfate are required for HSV infection. If so, they must be highly conserved and ubiquitous molecules to account for the very broad host range of HSV (fish, fowl, mammal).

gC is Principally Responsible for the Adsorption of HSV to Cells

The evidence that gC mediates the binding of HSV to cells includes the demonstration that gC has affinity for heparin and that gC-negative virions are impaired for adsorption (Herold *et al.*, 1991).

The results of affinity chromatography experiments done with solubilized virion envelope proteins showed that two of the HSV glycoproteins (gB and gC) can bind to heparin-Sepharose. Other experiments done with extracts of mutant virions indicated that the binding of gB was independent of the presence of gC and vice versa.

It had previously been shown that mutant virions devoid of gB could bind to cells as efficiently as wild-type virions (Cai *et al.*, 1988). We confirmed this finding. Similar experiments done with gC-negative mutant virions, however, demonstrated that the absence of gC was associated with significant impairment of adsorption (Fig. 3).

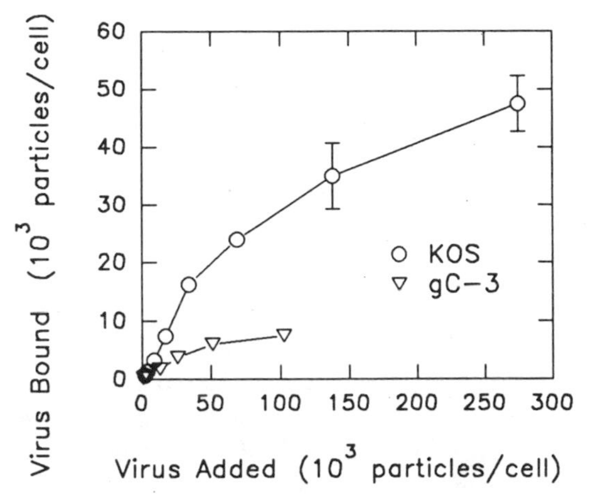

Fig. 3. Adsorption of HSV-1(strain KOS) and the gC-negative mutant HSV-1(KOS)gC-3 (Holland *et al.*, 1984) to human HEp-2 cells. Virions labeled with [3H]thymidine were purified and particle numbers were determined by electron microscopy. Cells plated on the bottoms of glass scintillation vials were exposed to various concentrations of virions at 4°C for 5 hrs. After washing away unbound virions, scintillation fluid was added to the vials for the quantitation of radioactivity bound to the cells. The specific infectivity of the KOS virions was significantly higher than that of the gC-3 virions (1 PFU per 60 particles of KOS vs 1 PFU per 830 particles of gC-3). Adapted from a figure published by Herold *et al.* (1991).

The reduction in binding efficiency did not appear to account fully for the reduced specific infectivity of the gC-negative virions (see legend to Fig. 3). We found that the gC-negative mutants were also impaired in penetration, in that there was a longer lag between adsorption and penetration for the mutant than for wild-type virus (Fig. 4). These defects in adsorption and penetration observed for the gC-negative mutant were unexpected inasmuch as such mutants are readily propagated in cell culture and do not appear to be defective unless these specific tests are applied.

Although gC is clearly important for the adsorption of wild-type virus to cells, gC-negative mutants can bind to cells and initiate infection, albeit inefficiently. The question arises as to the identity of cell surface receptors used by the mutant virus and of the virion glycoprotein(s) that mediate adsorption. It appears that cell surface heparan sulfate is required for the adsorption of gC-negative virus as well as wild-type virus. Mutant virus fails

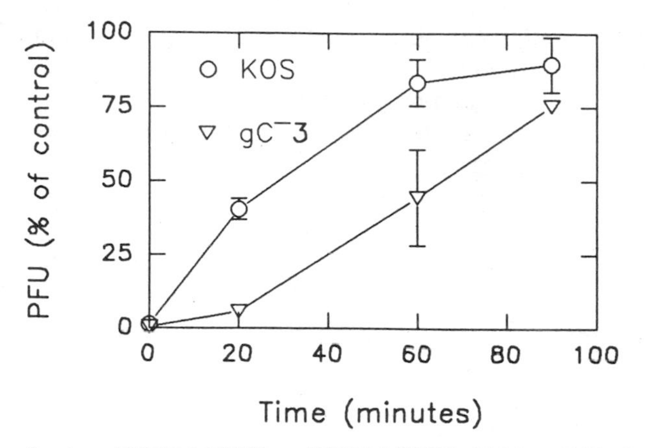

Fig. 4. Rate of entry of HSV-1(KOS) and HSV-1(KOS)gC-3 into HEp-2 cells. HEp-2 cells were exposed to purified virus for 2 hrs at 4°C and then washed to remove unbound virus. At the times indicated after raising the temperature to 37°C, the cells were exposed briefly to citrate buffer (pH 3.0) to inactivate virus that had adsorbed to the cells but not yet penetrated. This treatment has no effect on the ability of cells, into which virus has already penetrated, to initiate plaque formation. The cells were then washed with medium at physiological pH and incubated for 3 days to allow plaques to develop. The numbers of surviving plaque-forming units (PFU), expressed as the percentage of the number obtained after washing the cells with buffer at pH 7.4, are shown. Adapted from a figure published by Herold *et al.* (1991).

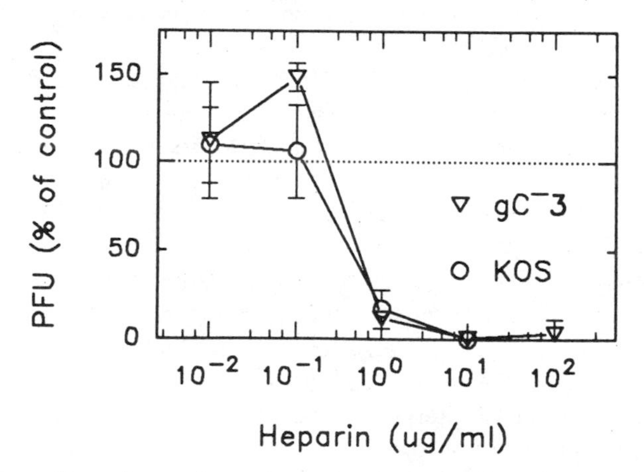

Fig. 5. Effect of heparin on plaque formation by HSV-1(KOS) and HSV-1(KOS)gC-3. HEp-2 cells were exposed, for 2 hrs at 37°C, to virus inocula containing the concentrations of heparin indicated. The cells were then washed and incubated for 3 days to allow plaques to develop. The numbers of PFU detected in the presence of heparin, expressed as a percentage of the number detected in the absence of heparin, are shown. Adapted from a figure published by Herold *et al.* (1991).

```
                                             pl pa    t     k   n
gB-1 (17)     apsspgtp---gvaaatqaanggpatpappapgapptgdpKpKKnRKpKp - 77
              ::..: .:   :  ::: :::::::.  :: :    :   : :  . ::
gB-2 (333)    apaapaapRasggvaatvanggpasRpppvpspattKaRKRKtKKppKRp - 72

              p
              t
gB-1 (17)     pKppRpagdnatvaaghatlRehlRdiKaentdanfyvcppptgapvvqf -127
              :   :  ::::::::::: :::.:: ::.::.::::::::::::::::
gB-2 (333)    peatpppdanatvaaghatlRahlReiKvenadaqfyvcppptgapvvqf -122

-------------------------------------------------------------------

gC-1 (17)     mapgRvglavvlwsllwlgagvsggsetastgptitagavknaseaptsg - 50
              :: ::::::: :: :::.:  :       :: : ::: :   ::: :  :
gC-2 (333)    malgRvglavglwgllwvgvvvvlan--aspgRtitvgpRgnasnaapsa - 48

gC-1 (17)     spgsaaspevtptstpnpnnvtqnKttptepasppttpKptstpKsppts -100
              ::  :..:  :::  :.:   :  :.. . :: ::
gC-2 (333)    spRnasapRttptp-pqpRKatKsKastaKpappp-------------- - 82

gC-1 (17)     tpdpKpKnnttpaKsgRptKppgpvwcdRRdplaRygsRvqiRcRfRnst -150
                            :.: :    :: : : ::::::::::::::: :::
gC-2 (333)    ------------KtgppKtssepvRcnRhdplaRygsRvqiRcRfpnst -119

gC-1 (17)     RmefRlqiwRysmgpsppiapapdleevltnitappggllvydsapnltd -200
              : : :::::::.      :  :: ::::. :..::::: ::::::: ::
gC-2 (333)    RtesRlqiwRyatatdaeigtapsleevmvnvsappggqlvydsapnRtd -169
```

Fig. 6. Amino acid sequences near the N-termini of gB and gC. Numbering is from the first amino acid of the translation product including the signal sequence. For gB the first amino acid shown is the first amino acid of the mature protein after cleavage of the signal sequence. Alignments of the HSV-1(strain 17) and HSV-2(strain 333) forms of each protein (e.g., gB-1 and gB-2) were done using the PALIGN program in the PCGENE suite. The double dots indicate sequence identity and the single dots, sequence similarity; the dashes represent gaps introduced to maximize alignment. Analysis of these sequences and others in Version 13 of the Swiss Protein Database revealed that, in the regions shown here, gB-2 sequences for two strains were identical as were gC-1 sequences and gC-2 sequences for two strains each. gB-1 sequences for four strains revealed the sequence variations indicated by the substitutions shown above the line of strain 17 sequence. The basic amino acids are shown in underlined capital letters.

to bind to cells devoid of heparan sulfate. Moreover, heparin inhibits the binding of gC-negative virus to cells and infection of the cells as effectively as it inhibits wild-type virus (Fig. 5). Because gB is also a heparin-binding protein, it seems likely that gB mediates the binding of gC-negative virions to cells.

The structural features of gB and gC necessary for binding to heparin or to cell surface heparan sulfate have not yet been identified. Although functional interactions between proteins and GAGs are not necessarily exlusively electrostatic, relatively high local concentrations of basic amino acids are characteristic of the heparin-binding domains of some proteins (Cardin *et al.*, 1989). Fig. 6 shows the amino acid sequences at the N-termini of the HSV-1 and HSV-2 forms of both gB and gC. There is a short Lys-rich region in gB and a longer Lys- and Arg-rich region in gC. It is of interest that the amino acid sequences of gB and gC are highly conserved between serotypes except at the N-termini and particularly in the basic regions indicated. Moreover, variations in amino acid sequence of gB for different strains of one serotype are concentrated mostly in and near the Lys-rich region (too few gC sequences have been determined to permit a similar assessment of strain variation).

If the basic regions shown in Fig. 6 prove to mediate the binding of gB and gC to heparan sulfate or heparin, the possibility exists that the observed variability in amino acid sequences may be associated with variability in specificity of recognition of heparan sulfate. In other words, gB and gC probably recognize different structural features of heparan sulfate. The different forms of gB (and gC) may also recognize somewhat different structural features of heparan sulfate.

Comparisons of HSV with Other Members of the Herpesvirus Family

The different human herpesviruses are biologically quite diverse and include Epstein-Barr virus (EBV), cytomegalovirus, varicella-zoster virus, human herpesviruses 6 and 7, as well as HSV. The host ranges, tissue tropisms and pathogenicity of these viruses are so different that similarities in mechanisms of infectivity are not necessarily to be expected. Available evidence supports this notion. The only known cell surface receptor for EBV is CR2, the C3d receptor (Fingeroth *et al.*, 1984). The distribution of this receptor, principally on B lymphocytes and perhaps on some epithelial cells, explains the limited host cell range of the virus. The EBV glycoprotein designated gp350 mediates the binding of virus to CR2 (Nemerow *et al.*, 1987; Tanner *et al.*, 1987). Interestingly, gp350 binds to a complement receptor and gC, the HSV glycoprotein responsible for virus adsorption, binds to a complement component C3b (Friedman *et al.*, 1984; Eisenberg *et al.*, 1987; McNearney *et al.*, 1987). Although gp350 and gC show little, if any, homology in primary amino acid sequence, the fact that they both interact with the complement system suggests a possible evolutionary relationship. If these two glycoproteins have a common ancestor, they have evolved to recognize quite different kinds of cell surface molecules as receptors for viral infection.

Animal herpesviruses that are biologically similar to one or another of the human viruses have been described. For example, pseudorabies virus (PRV) of pigs is much more similar to HSV in its biology and pathogenicity than are any of the other human herpesviruses. Recently, it has been shown that heparan sulfate also serves as cell surface receptor for PRV (Mettenleiter *et al.*, 1990). The PRV glycoprotein that mediates the adsorption of virus to cells is gIII, which has heparin-binding activity and is the homolog of HSV gC (Robbins *et al.*, 1986a; Robbins *et al.*, 1986b; Ben-Porat *et al.*, 1986; Schreurs *et al.*, 1988; Zuckermann *et al.*, 1989; Mettenleiter, 1989; Mettenleiter *et al.*, 1990). Moreover, gIII-negative PRV mutants, similar to gC-negative HSV mutants, have impaired infectivity but can be propagated in cell culture. Although these similarities between HSV and PRV are striking, one difference has been noted. Cell surface heparan sulfate apparently does not serve as receptor for the adsorption of gIII-negative PRV mutants to cells (Mettenleiter *et al.*, 1990). It is perhaps significant that the PRV homolog of HSV gB (gII) does not have the Lys-rich region that is characteristic of gB (Robbins *et al.*, 1987). None of the gB homologs specified by any of the human herpesviruses have this Lys-rich region either, despite the fact that gB is the most highly conserved of all the herpesvirus glycoproteins.

Summary

There are a number of viruses for which the only known specific interaction with the cell surface is binding to a carbohydrate moiety of a cell surface glycoprotein or glycolipid. In our view, HSV belongs in this category, at least for the present. The possibility exists, for any of these viruses, that the initial specific interaction with a carbohydrate leads to other specific interactions with proteins, lipids, etc. Because most of these viruses have broad host ranges, any specific receptors required must be highly conserved and broadly distributed.

We envision that the interaction of HSV with heparan sulfate and heparin-like molecules could have functional significance in several ways. First, the binding of HSV to cell surface heparan sulfate serves to concentrate virus on the cell surface. Second, inasmuch as many heparin-binding proteins are altered structurally and functionally by their interactions with heparin, the possibility exists that the binding of HSV glycoproteins to heparan sulfate activates functions required for viral penetration. Third, because both heparin and heparin-binding proteins can inhibit HSV infection and because structural tissue elements such as basement membranes are rich in heparan sulfate proteoglycans, the possibility exists that interactions of virus with any of these non-cell-associated molecules may serve to limit the spread of HSV infection, explaining perhaps the localized nature of most HSV infections.

References

Baird, A., Florkiewicz, R.Z., Maher, P.A., *et al.*, 1990, Mediation of virion penetration into vascular cells by association of basic fibroblast growth factor with Herpes simplex virus type 1, *Nature*. 348:344-346.

Bame, K.J. and Esko, J.D., 1989, Undersulfated heparan sulfate in a Chinese hamster ovary cell mutant defective in heparan sulfate N-sulfotransferase, *J. Biol. Chem.* 264:8059-8065.

Ben-Porat, T., DeMarchi, J.M., Lomniczi, B., *et al.*, 1986, Role of glycoproteins of pseudorabies virus in eliciting neutralizing antibodies, *Virology* 154:325-334.

Burgess, W.H. and Maciag, T., 1989, The heparin-binding (fibroblast) growth factor family of proteins, *Annu. Rev. Biochem.* 58:575-606.

Cai, W.H., Gu, B., and Person, S., 1988, Role of glycoprotein B of Herpes simplex virus type 1 in viral entry and cell fusion, *J. Virol.* 62:2596-2604.

Campadelli Fiume, G., Qi, S., Avitabile, E., *et al.*, 1990, Glycoprotein D of Herpes simplex virus encodes a domain which precludes penetration of cells expressing the glycoprotein by superinfecting Herpes simplex virus, *J. Virol.* 64:6070-6079.

Campadelli-Fiume, G., Arsenakis, M., Farabegoli, F., *et al.*, 1988, Entry of Herpes simplex virus 1 in BJ cells that constitutively express viral glycoprotein D is by endocytosis and results in degradation of the virus, *J. Virol.* 62:159-167.

Cardin, A.D. and Weintraub, H.J.R., 1989, Molecular modeling of protein-glycosaminoglycan interactions, *Arteriosclerosis* 9:21-32.

Corey, L. and Spear, P.G., 1988, Infections with Herpes simplex viruses, N. Engl. J. Med. 314:686-689; 749-757.

Desai, P.J., Schaffer, P.A., and Minson, A.C., 1988, Excretion of non-infectious virus particles lacking glycoprotein H by a temperature-sensitive mutant of Herpes simplex virus type 1: evidence that gH is essential for virion infectivity, *J. Gen. Virol.* 69:1147-1156.

Eisenberg, R.J., Ponce de Leon, M., Friedman, H.M., *et al.*, 1987, Complement component C3b binds directly to purified glycoprotein C of Herpes simplex virus types 1 and 2, *Microb. Pathog.* 3:423-435.

Esko, J.D., Rostand, K.S. and Weinke, J.L., 1988, tumor formation dependent on proteoglycan biosynthesis, *Science* 241: 1092-1096.

Esko, J.D., Stewart, T.E., and Taylor, W.H., 1985, Animal cell mutants defective in glycosaminoglycan biosynthesis, *Proc. Natl. Acad. Sci. USA* 82:3197-3201.

Esko, J.D., Weinke, J.L., Taylor, W.H., *et al.*, 1987, Inhibition of chondroitin and heparan sulfate biosynthesis in Chinese hamster ovary cell mutants defective in galactosyltransferase I, *J. Biol. Chem.* 262:12189-12195.

Fingeroth, J.D., Weis, J.J., Tedder, T.F., *et al.*, 1984, Epstein-Barr virus receptor of human B lymphocytes is the C3d receptor CR2, *Proc. Natl. Acad. Sci. USA* 81:4510-4514.

Friedman, H.M., Cohen, G.H., Eisenberg, R.J., *et al.*, 1984, Glycoprotein C of Herpes simplex virus 1 acts as a receptor for the C3b complement component on infected cells, *Nature* 309:633-635.

Fuller, A.O. and Spear, P.G., 1987, Anti-glycoprotein D antibodies that permit adsorption but block infection by Herpes simplex virus 1 prevent virion-cell fusion at the cell surface, *Proc. Natl. Acad. Sci. USA* 84:5454-5458.

Fuller, A.O., Santos, R.E., and Spear, P.G., 1989, Neutralizing antibodies specific for glycoprotein H of Herpes simplex virus permit viral attachment to cells but prevent penetration, *J. Virol.* 63:3435-3443.

Herold, B.C., WuDunn, D., Soltys, N., *et al.*, 1991, Glycoprotein C of Herpes simplex virus type 1 plays a principal role in the adsorption of virus to cells and in infectivity, *J. Virol.* 65:1090-1098.

Highlander, S.L., Sutherland, S.L., Gage, P.J., *et al.*, 1987, Neutralizing monoclonal antibodies specific for Herpes simplex virus glycoprotein D inhibit virus penetration, *J. Virol.* 61:3356-3364.

Highlander, S.L., Cai, W.H., Person, S., *et al.*, 1988, Monoclonal antibodies define a domain on Herpes simplex virus glycoprotein B involved in virus penetration, *J. Virol.* 62:1881-1888.

Holland, T.C., Homa, F.L., Marlin, S.D., *et al.*, 1984, Herpes simplex virus type 1 glycoprotein C-negative mutants exhibit multiple phenotypes, including secretion of truncated glycoproteins, *J. Virol.* 52:566-574.

Hook, M., Kjellen, L., Johansson, S., *et al.*, 1984, Cell-surface glycosaminoglycans, *Ann. Rev. Biochem.* 53:847-869.

Johnson, D.C. and Ligas, M.W., 1988, Herpes simplex viruses lacking glycoprotein D are unable to inhibit virus penetration: quantitative evidence for virus-specific cell surface receptors, *J. Virol.* 62:4605-4612.

Johnson, D.C., Burke, R.L., and Gregory, T., 1990, Soluble forms of Herpes simplex virus glycoprotein D bind to a limited number of cell surface receptors and inhibit virus entry into cells, *J. Virol.* 64:2569-2576.

Johnson, R.M. and Spear, P.G., 1989, Herpes simplex virus glycoprotein D mediates interference with Herpes simplex virus infection, *J. Virol.* 63:819-827.

Kaner, R.J., Baird, A., Mansukhani, A., *et al.*, 1990, Fibroblast growth factor receptor is a portal of cellular entry for Herpes simplex virus type 1, *Science.* 248:1410-1413.

Ligas, M.W. and Johnson, D.C., 1988, A Herpes simplex virus mutant in which glycoprotein D sequences are replaced by *B*-galactosidase sequences binds to but is unable to penetrate into cells, *J. Virol.* 62:1486-1494.

Mansukhani, A., Moscatelli, D., Talarico, D., *et al.*, 1990, A murine fibroblast growth factor (FGF) receptor expressed in CHO cells is activated by basic FGF and Kaposi FGF, *Proc. Natl. Acad. Sci. USA* 87:4378-4382.

McNearney, T.A., Odell, C., Holers, V.M., *et al.*, 1987, Herpes simplex virus glycoproteins gC-1 and gC-2 bind to the third component of complement and provide protection against complement-mediated neutralization of viral infectivity, *J. Exp. Med.* 166:1525-1535.

Mettenleiter, T.C., 1989, Glycoprotein gIII deletion mutants of pseudorabies virus are impaired in virus entry, *Virology*. 171:623-625.

Mettenleiter, T.C., Zsak, L., Zuckermann, F., *et al.*, 1990, Interaction of glycoprotein gIII with a cellular heparinlike substance mediates adsorption of pseudorabies virus, *J. Virol.* 64:278-286.

Nahmias, A.J. and Kibrick, S., 1964, Inhibitory effect of heparin on Herpes simplex virus, *J. Bacteriology* 87:1060-1066.

Nemerow, G.R., Mold, C., Schwend, V.K., *et al.*, 1987, Identification of gp350 as the viral glycoprotein mediating attachment of Epstein-Barr virus (EBV) to the EBV/C3d receptor of B cells: sequence homology of gp350 and C3 complement fragment C3d, *J. Virol.* 61:1416-1420.

Olwin, B.B. and Hauschka, S.D., 1989, Cell type and tissue distribution of the fibroblast growth factor receptor, *J. Cell. Biochem*. 39:443-454.

Robbins, A.K., Watson, R.J., Whealy, M.E., *et al.*, 1986a, Characterization of a pseudorabies virus glycoprotein gene with homology to Herpes simplex virus type 1 and type 2 glycoprotein C, *J. Virol.* 58:339-347.

Robbins, A.K., Whealy, M.E., Watson, R.J., *et al.*, 1986b, Pseudorabies virus gene encoding glycoprotein gIII is not essential for growth in tissue culture, *J. Virol.* 59:635-645.

Robbins, A.K., Dorney, D.J., Wathen, M.W., *et al.*, 1987, The pseudorabies virus gII gene is closely related to the gB glycoprotein gene of Herpes simplex virus, *J. Virol.* 61:2691-2701.

Roizman, B. and Sears, A., 1990, Herpes simplex viruses and their replication, in Virology (Fields, B.N., Ed.) 2nd edition, Raven Press,Ltd., New York. 1795-1841.

Sarmiento, M., Haffey, M., and Spear, P.G., 1979, Membrane proteins specified by Herpes simplex viruses. III. Role of glycoprotein VP7 (B2) in virion infectivity, *J. Virol.* 29:1149-1158.

Schreurs, C., Mettenleiter, T.C., Zuckermann, F., *et al.*, 1988, Glycoprotein gIII of pseudorabies virus is multifunctional, *J. Virol.* 62:2251-2257.

Shieh, M.-T. and Spear, P.G., 1991, Fibroblast growth receptor: Does it have a role in the binding of Herpes simplex virus? *Science* 253: 208-210.

Stannard, L.M., Fuller, A.O., and Spear, P.G., 1987, Herpes simplex virus glycoproteins associated with different morphological entities projecting from the virion envelope, *J. Gen. Virol.* 68:715-725.

Takemoto, K.K. and Fabisch, P., 1964, Inhibition of Herpes simplex virus by natural and synthetic acid polysaccharides, *Proc. Soc. Exp. Biol. Med.* 116:140-144.

Tanner, J., Weis, J., Fearon, D., *et al.*, 1987, Epstein-Barr virus gp350/220 binding to the B lymphocyte C3d receptor mediates adsorption, capping and endocytosis, *Cell* 50:203-213.

Vaheri, A., 1964, Heparin and related polyionic substances as virus inhibitors, *Acta Path. Microbiol. Scand. Suppl.* 171:7-97.

Wittels, M. and Spear, P.G., 1991, Penetration of cells by Herpes simplex virus does not require a low pH-dependent endocytic pathway, *Virus Res.*, 18: 271-290.

WuDunn, D. and Spear, P.G., 1989, Initial interaction of Herpes simplex virus with cells is binding to heparan sulfate, *J. Virol.* 63:52-58.

Zuckermann, F., Zsak, L., Reilly, L., *et al.*, 1989, Early interactions of pseudorabies virus with host cells: functions of glycoprotein gIII, *J. Virol.* 63:3323-3329.

Interaction of Poliovirus with its Immunoglobulin-like Cell Receptor

Vincent R. Racaniello, Eric G. Moss, Gerardo Kaplan, and Ruibao Ren

Introduction

Understanding the pathogenesis of virus infections requires study of both viral and cell factors that contribute to the outcome of an infection. Our work has addressed two aspects of poliovirus pathogenesis: the molecular and functional basis of host range and tissue tropism. It has become apparent that understanding virus-receptor interactions is necessary to provide explanations of both characteristics. Therefore, studies on the cellular receptor for poliovirus (PVR) have become an integral part of our research aims. We shall, in this communication, first consider briefly the structure of the PVR and some of its properties, and then we will turn to studies on the role of the PVR in poliovirus host range and tissue tropism.

Poliovirus is an RNA containing virus with a protein shell of icosahedral symmetry. The viral capsid is comprised of 60 copies each of proteins VP1, VP2, VP3 and VP4. The atomic structure of the virion has been solved, enabling structure-function studies of exquisite specificity (Hogle et al., 1985). The virus initiates infection by binding to a cell receptor that is a novel member of the immunoglobulin superfamily of proteins (Mendelsohn *et al.*, 1989). The PVR is a transmembrane glycoprotein consisting of three immunoglobulin-like domains, a transmembrane segment and a cytoplasmic tail. The natural function of the PVR in cells is not known, but it is likely to be involved in ligand binding or perhaps cell-cell recognition.

Alternatively Spliced Forms of the PVR

Our initial studies revealed that HeLa cells contain two major receptor mRNAs which encode proteins with two different cytoplasmic tails (Mendelsohn *et al.*, 1989). Both PVR isoforms can serve as poliovirus receptors. Recently we identified two additional PVR mRNAs in HeLa cells that lack sequences encoding the transmembrane domain. The two mRNAs encode proteins that lack amino acids 331-385 and 339-385, respectively.

The two deleted forms are most likely derived by alternative splicing of the mRNA PVR gene, as they are found in mouse L cells that have been transformed with the PVR gene. The function of these deleted forms, which are presumably secreted from the cell, is not known, but their presence raises the possibility that the PVR may serve as ligand as well as receptor.

Neutralization of Poliovirus with Soluble PVR

One of our goals is to obtain a detailed picture of the poliovirus-PVR interaction. To provide sufficient material for such studies, the PVR was expressed in insect cells, using baculovirus vectors (Kaplan *et al.*, 1990a). The PVR was synthesized in insect cells as a 67 kd glycoprotein that is inserted into the cell membrane and is able to bind poliovirus. Poliovirus infectivity was neutralized by incubation of virus with PVR that was solubilized from the insect cells. Neutralization of viral infectivity resulted from PVR-mediated conversion of native poliovirions to altered particles. These particles are known to result from incubation of poliovirus with susceptible cells at 37 C, and differ significantly from native virions in a number of properties, including sedimentation coefficient (135S), increased hydrophobicity, lack of VP4 and lack of infectivity (Fricks and Hogle, 1990). The virus must undergo significant conformational transitions to become altered particles. The ability to produce large quantities of altered particles, using soluble PVR, now enables resolution of the crystallographic structure of these particles. This information will provide information on the precise structural changes that accompany the transition to altered particles.

It has been suggested that altered particles might be intermediates in poliovirus entry (Fricks and Hogle, 1990). Our results indicate that conversion of poliovirus to altered particles may occur by interaction with the PVR at neutral pH, in the absence of cell membranes. These results suggest that poliovirus particles may become altered when they first interact with the PVR at the cell surface. The altered particles, which are hydrophobic, might then be taken into the cell by endocytosis. Uncoating of the viral RNA must then occur, although the requirements for this step are not known. Alternatively, the altered virions bound to the cell surface might directly discharge the viral RNA into the cytoplasm. The altered particles might, for example, induce formation of a pore or channel in the cell membrane, through which the viral RNA might enter into the cytoplasm. Although it has been suggested that poliovirus enters the cell by receptor-mediated endocytosis (Madshus *et al.*, 1984), we recently found that deletion of the cytoplasmic tail from the PVR does not interfere with its ability to support poliovirus infection. Removal of the cytoplasmic tail from some membrane proteins interferes with their uptake into the cell by endocytosis (Maddon *et al.*, 1988). Our result suggests but does not prove that poliovirus might not require endocytosis to productively infect a cell.

Neutralization of poliovirus with soluble PVR, derived from insect cells, is not always complete. Despite using increasing levels of solubilized PVR, residual non-neutralized poliovirus is always observed (Kaplan *et al.*, 1990b). To determine whether the residual infectivity was due to inefficient neutralization or viral variants resistant to PVR-mediated neutralization, virus was neutralized with soluble PVR and viruses that escaped neutralization were plaque purified. After several rounds of neutralization and plaque purification, five isolates were obtained that were relatively more resistant to neutralization with soluble PVR than wild type virus. One of the five isolates was completely resistant to neutralization, even at high levels of soluble PVR, while the other isolates were neutralized only with high amounts of PVR. These mutants have been termed soluble receptor-resistant (srr) mutants. The srr mutants show no obvious phenotypes in cell culture, as determined by plaque assay and one-step growth analysis.

Analysis of the binding of srr mutants to HeLa cells and isolated HeLa cell membranes indicated that they bind less efficiently than wild type virus. The resistance of the srr mutants to PVR-mediated neutralization may therefore result from their reduced ability to bind the soluble material. However, the srr mutants must retain sufficient binding capacity to enable normal replication in cultured cells. These results have two implications. First, caution must be exercised when considering the use of soluble cell receptors as antiviral agents (Marlin *et al.*, 1990). Our results indicate that it is possible to select for viral mutants that are resistant to receptor-mediated neutralization and which might cause disease. Second, it will be of interest to identify the amino acid changes that lead to resistance to neutralization with soluble PVR. The location of such residues on the three-dimensional structure of the virion might provide information about amino acids involved in contact with the cell receptor.

Determinants of Poliovirus Host Range

Poliovirus strains have distinct host range limitations. Some strains, such as P1/Mahoney, cause disease only in primates, while other strains, such as P2/Lansing, can cause paralysis in mice as well. One of our central interests is to understand the molecular and functional basis of poliovirus host range. Several years ago we began to address this question by studying the P2/Lansing strain of poliovirus, which causes poliomyelitis in mice after intracerebral inoculation (Armstrong, 1939). Viral recombinants between P2/Lansing and the mouse-avirulent P1/Mahoney were initially used to determine that the ability of P2/Lansing to infect mice mapped to the capsid proteins (La Monica *et al.*, 1986). Analysis of the mouse neurovirulence of P2/Lansing variants resistant to neutralization with monoclonal antibodies implicated neutralization antigenic site 1 (N-Ag1) in host range. Finally, the ability to infect mice was mapped to an eight amino acid sequence of P2/Lansing capsid protein VP1 corresponding to N-Ag1 (Martin *et al.*, 1988;

Murray *et al.*, 1988). Substitution of amino acids 95-102 of P1/Mahoney with the corresponding sequence from P2/Lansing produced a chimeric virus (W1/2-1D-1) that was neurovirulent in mice.

A simple explanation for the ability of P2/Lansing, as well as the chimeric virus W1/2-1D-1, to cause disease in mice is that these viruses are able to recognize a receptor on the surface of mouse neurons that P1/Mahoney cannot. The location of amino acids 95-102, also known as the BC loop, on the virion surface is consistent with this notion. Five copies of the BC loop encircle the fivefold axis of symmetry, and form a prominent protrusion on the virion surface (Hogle *et al.*, 1985). This protrusion could easily engage with a host cell receptor. However, because we have found that the BC loop is nonessential for growth in cultured primate cells, this suggestion carries the implication that the recognition site used by the virus in cultured primate cells differs from that used in mice.

Our recent findings have lead to a modification of our hypothesis on the functional basis of poliovirus host range. We have found that substitution of the BC loop of P2/Lansing with the sequence from P1/Mahoney produced a virus, called R2-1D-3, that was highly attenuated in mice, with an LD_{50} of $>1.4 \times 10^9$. To provide information on which residues of the BC loop are important in mouse neurovirulence, we thought it might be useful to select revertants of higher neurovirulence. Virus was therefore recovered from the spinal cords of paralyzed mice that had been inoculated with R2-1D-3 and passaged two additional times in mice. The resulting spinal cord isolate, scp3, was 1000 times more neurovirulent in mice than the parental R2-1D-3. Sequence analysis of scp3 indicated that there were no amino acid changes within the BC loop. Rather, scp3 contained a second site mutation at amino acid 40 of VP1 (Glu to Gly) that was shown to be responsible for restoring neurovirulence to R2-1D-3. Similar analysis was conducted with another attenuated virus, R2-1D-6, in which amino acids 95-97 of the P2/Lansing B-C loop were substituted with the sequence from P1/Mahoney. Passage of this virus in mice resulted in a variant, scp6, that was approximately 100 times more virulent than R2-1D-6. Analysis of the sequence changes in scp6 indicated that a second site mutation at amino acid 54 of VP1, that resulted in a change from Pro to Ser, was responsible for the increased neurovirulence of the variant.

In two independent isolates, a second site mutation in the N-terminus of VP1 was responsible for restoring neurovirulence to a virus that had been attenuated by replacement of the B-C loop. The location of these second site suppressors in the virion structure, however, was a complete surprise. Residues 40 and 54 of VP1 are located on the interior of the virion, very distant from the BC loop. The location of these suppressor mutations provides clues to the functional basis of poliovirus host range in mice. There are two possibilities for the mechanism of action of the suppressor mutations. In the first scenario, the BC loop variants R2-1D-3 and -6 might be attenuated due to a defect in recognizing the mouse cell receptor. The second-site suppressors might restore neurovirulence to these strains by changing the conformation of the BC loop, enabling it to recognize the murine receptor. However, this explanation seems unlikely, given the great distance between the suppressors and the BC loop.

A different but more likely possibility is that the suppressors might influence the ability of the capsid to undergo the conformational transitions required for initiation of infection. The suppressors are in fact located near sites that are thought to regulate conformational transitions in the virion (Filman *et al.*, 1989). The suppressors might therefore restore to the attenuated viruses R2-1D-3 and -6 the ability to undergo conformational transitions after binding a mouse cell receptor. An extension of this argument is that such alterations are also conferred by the P2/Lansing BC loop, but not by the Mahoney BC loop. Additional study of the virus-receptor interaction will be required to determine whether host range is determined by regulation of conformational transitions of the virion.

Role of the PVR in Poliovirus Tissue Tropism

The restricted cell and tissue tropism of poliovirus infection in primates has been well described (Bodian, 1959). Poliovirus infection begins when virus is ingested and replicates in the gut, leading to a viremia. In a small number of infections virus spreads to the central nervous system, where it replicates nearly exclusively in motor neurons, leading the the characteristic flaccid paralysis. We have been interested in determining the mechanisms that limit poliovirus infection to only a few cell types, despite its access to a wide variety of tissues. Many years ago it was proposed that the PVR is the determinant of poliovirus tissue tropism (Holland, 1961).

When molecular clones of PVR cDNA were isolated (Mendelsohn *et al.*, 1989) it became possible to readdress the relationship between PVR expression and poliovirus tissue tropism using PVR-specific nucleic acid and antibody probes. PVR RNA and protein was found to be expressed in a wide variety of human tissues, including those that are not infected by poliovirus, such as kidney (Mendelsohn *et al.*, 1989; Freistadt *et al.*, 1990). These findings suggested that poliovirus tissue tropism is controlled either by modifications of the receptor or the availability of factors required for productive infection.

To circumvent the difficulties associated with studying the relationship between PVR expression and tissue tropism in humans, we have established a transgenic mouse model for poliomyelitis (Ren *et al.*, 1990). The human PVR gene was isolated and used to establish several lines of transgenic mice. The PVR transgenic mice express PVR mRNA and poliovirus binding sites in a wide variety of mouse tissues, including brain, spinal cord, lung, liver, and kidney. When inoculated intracerebrally with poliovirus, PVR transgenic mice develop paralytic disease that clinically and histopathologically resembles human poliomyelitis. PVR transgenic mice are susceptible to infection with all three serotypes of poliovirus, including P1/Mahoney, which cannot infect normal mice. Transgenic mice expressing human PVR are therefore a new model for studying poliomyelitis.

In view of the widespread expression of PVR mRNA and poliovirus virus binding sites in transgenic mouse tissues, it was of interest to determine the sites of poliovirus replication in these animals. After intracerebral inoculation of PVR transgenic mice, poliovirus replicated to high titers in brain and spinal cord. However, after intravenous or intra-peritoneal inoculation, poliovirus replication was not detected in non-neural tissues. An example of such an experiment is shown in Figure 1.

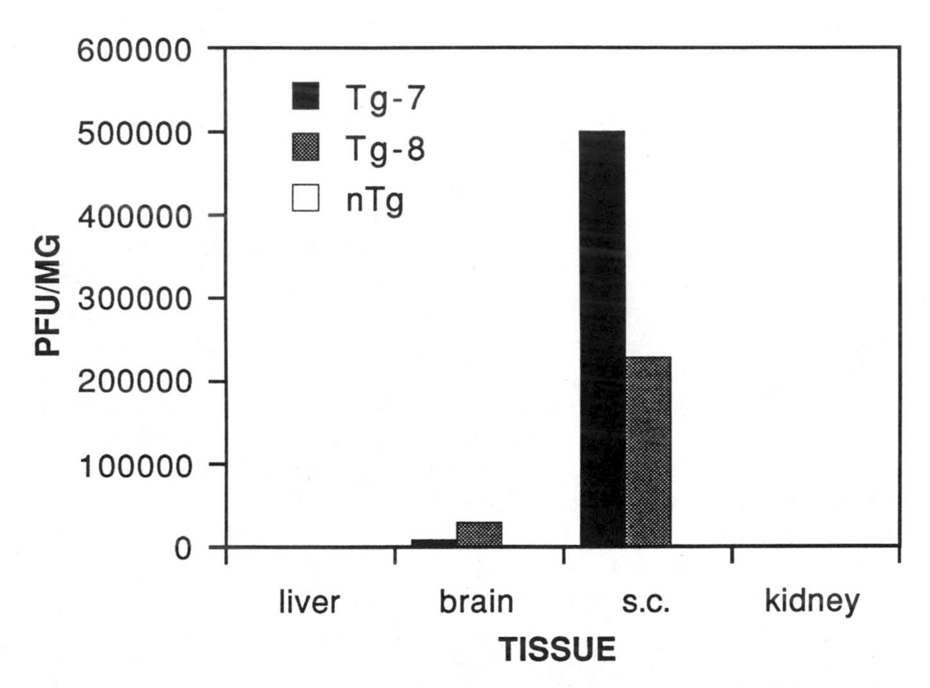

Figure 1. Poliovirus replication in transgenic mouse tissues. PVR transgenic mice were inoculated with 1×10^9 pfu of P1/Mahoney. Virus titers in liver, brain, spinal cord and kidney were determined on day 7 and 8 (Tg-7, -8). nTg, nontransgenic mice.

In this experiment, mice were inoculated intravenously with 1×10^9 pfu of P1/Mahoney. At seven and eight days after infection, a paralyzed animal was sacrificed and virus titer determined in liver, brain, spinal cord and kidney. The results indicate that virus only replicates in brain and spinal cord. In a similar experiment, PVR transgenic mice were inoculated intraperitoneally, and virus titer in organs was determined daily up to day 8 post infection. No evidence for viral replication was observed in any tissues other than brain and spinal cord.

These results suggest that despite widespread expression of the human PVR in transgenic mice, poliovirus replication is restricted to the brain and spinal cord. PVR transgenic mice can therefore be used to study the basis of poliovirus tissue tropism. Work is currently under way to determine why poliovirus is unable to replicate in specific organs, such as kidney. As part of this study, PVR and poliovirus mRNA expression is being examined by *in situ* hybridization.

Early experiments suggested that poliovirus binding sites are mainly expressed in the gut, spinal cord and brain, although expression in liver and kidney was sometimes detected (Holland and McLaren, 1961; Kunin and Jordan, 1961). These results contrast with those obtained with transgenic mice, where poliovirus binding sites are found in all organs examined. We believe that the difference is caused by overexpression of the PVR in transgenic mice. The transgenic mice that we have examined carry ten copies of the PVR gene, and express high levels of PVR mRNA, especially in spinal cord and brain. If the PVR is labile in tissue homogenates, it might not be detected unless it were expressed at high levels. Thus, the ability to detect poliovirus binding sites in many transgenic mouse tissues may result from overexpression of the protein, while failure to detect binding sites in many human tissues may be a consequence of low expression of a labile protein.

Summary

Our studies on poliovirus pathogenesis have addressed the molecular and functional basis of poliovirus host range and tissue tropism. The molecular basis for the ability of some poliovirus strains to infect mice has been mapped to a small sequence of capsid protein VP1. However, the functional basis for host range is still not understood. Our results demonstrate that poliovirus host range in mice can be controlled in three ways: by the sequence of the BC loop, by the sequence at the N-terminus of VP1, and by expressing the human PVR in mice. Thus, host range is controlled by a virus receptor interaction, although the precise event involved is not yet known. The study of poliovirus host range in this way provides a means of directly investigating virus-receptor interactions.

The basis of poliovirus tissue tropism has not yet been elucidated. We have established a transgenic mouse model for poliomyelitis that will certainly be useful for probing the functional basis of restricted tropism. Despite the fact that the transgenic mice express virus-binding PVR in many tissues, poliovirus replication is restricted to the neuraxis. Clearly the next question to be addressed is why poliovirus cannot replicate in transgenic mouse tissues, despite expression of the PVR. An answer to this question should provide information on the basis of poliovirus tropism in humans as well.

By studying individual aspects of poliovirus pathogenesis in a mouse model, our hope is to provide a precise description of how poliovirus causes disease.

References

Armstrong, C., 1939, Successful transfer of the Lansing strain of poliomyelitis virus from the cotton rat to the white mouse, *Pub. Health Rep*. 54:2302-2305.

Bodian, D., 1959, Poliomyelitis: pathogenesis and histopathology. In Viral and Rickettsial Infections of Man, (T.M. Rivers and F.L. Horsfall, ed). Lippincott, Philadelphia, p. 479-498.

Filman, D.J., Syed, R., Chow, M., Macadam, A.J., Minor, P.D. and Hogle, J.M., 1989, Structural factors that control conformational transitions and serotype specificity in type 3 poliovirus, *EMBO J.* 8:1567-1579.

Freistadt, M.F., Kaplan, G. and Racaniello, V.R., 1990, Heterogeneous expression of poliovirus receptor-related proteins in human cells and tissues, *Mol. Cell Biol.* 10: 5700-5706.

Fricks, C.E. and Hogle, J.M., 1990, The cell-induced conformational change of poliovirus: Externalization of the amino terminus of VP1 is responsible for liposome binding, *J. Virol*. 64:1934-1945.

Hogle, J.M., Chow, M. and Filman, D.J., 1985, Three-dimensional structure of poliovirus at 2.9 A resolution, Science. 229:1358-1365.

Holland, J.J., 1961, Receptor affinities as major determinants of enterovirus tissue tropisms in humans, *Virology*. 15:312-326.

Holland, J.J. and McLaren, L.C., 1961, The location and nature of enterovirus receptors in susceptible cells, *J. Exp. Med*. 114:161-171.

Kaplan, G., Freistadt, M.S. and Racaniello, V.R., 1990a, Neutralization of poliovirus by cell receptors expressed in insect cells, *J. Virol.* 64:4697-4702.

Kaplan, G., Peters, D. and Racaniello, V.R., 1990b, Poliovirus mutants resistant to neutralization with soluble cell receptors, *Science*. 250: 1596-1599..

Kunin, C.M. and Jordan, W.S., 1961, *In vitro* adsorption of poliovirus by noncultured tissues. Effect of species, age and malignancy, *Am. J. Hyg*. 73:245-257.

La Monica, N., Meriam, C. and Racaniello, V.R., 1986, Mapping of sequences required for mouse neurovirulence of poliovirus type 2 Lansing, *J. Virol*. 57:515-525.

Maddon, P.J., McDougal, J.S., Clapham, P.R., Dalgleish, A.G., Jamal, S., Weiss, R.A. and Axel, R., 1988, HIV infection does not require endocytosis of its receptor, CD4, *Cell*. 54:865-874.

Madshus, I.H., Olsnes, S. and Sandvig, K., 1984, Mechanism of entry into the cytosol of poliovirus type 1: requirement for low pH, *J. Cell. Biol.* 98:1194-1200.

Marlin, S.D., Staunton, D.E., Springer, T.A., Stratowa, C., Sommergruber, W. and Merluzzi, V.J., 1990, A soluble form of intercellular adhesion molecule-1 inhibits rhinovirus infection, *Nature.* 344:70-72.

Martin, A., Wychowski, C., Couderc, T., Crainic, R., Hogle, J. and Girard, M., 1988, Engineering a poliovirus type 2 antigenic site on a type 1 capsid results in a chimaeric virus which is neurovirulent for mice, *EMBO J.* 7:2839-2847.

Mendelsohn, C., Wimmer, E. and Racaniello, V.R., 1989, Cellular receptor for poliovirus: Molecular cloning, nucleotide sequence and expression of a new member of the immunoglobulin superfamily, *Cell.* 56:855-865.

Murray, M.G., Bradley, J., Yang, X.F., Wimmer, E., Moss, E.G. and Racaniello, V.R., 1988, Poliovirus host range is determined by a short amino acid sequence in neutralization antigenic site I, *Science.* 241:213-215.

Ren, R., Costantini, F.C., Gorgacz, E.J., Lee, J.J. and Racaniello, V.R., 1990, Transgenic mice expressing a human poliovirus receptor: A new model for poliomyelitis, *Cell.* 63: 353-362.

Characterization and Expression of the Glycoprotein Receptor for Murine Coronavirus

Kathryn V. Holmes and Richard K. Williams

Introduction

Coronaviruses are a family of large, enveloped, viruses with helical nucleocapsids and a single stranded genomic RNA of approximately 30kb of message sense (Sturman and Holmes, 1983; Spaan *et al.*, 1988). These viruses cause respiratory and/or enteric infections in man and many domestic animals (Wege *et al.*, 1982). We are particularly interested in coronaviruses which infect the intestine. Pigs, cows, dogs, cats and mice are natural hosts for different coronaviruses each of which causes intestinal infection only in a single host species. Coronavirus infection results in a wide spectrum of clinical disease, ranging from inapparent infection of adult animals to severe, fatal neonatal diarrhea. Immunofluorescence studies show that in the intestine, coronavirus replication is limited to the epithelial cells, and multinucleated cells are frequently observed on the tips of the intestinal villi (Barthold *et al.*, 1982; Barthold *et al.*, 1985). We were interested in factors of the virus and the host which determine the species specificity and tissue tropism of these enterotropic coronaviruses.

As a model system to study the interactions of coronavirus virions with membranes of the normal target tissues from a naturally susceptible animal, we selected mouse hepatitis virus (MHV). MHV refers to a large group of serologically related coronaviruses which cause endemic and epidemic infections in mouse colonies (Lindsey, 1986). The tissue specificity of MHV strains is broader than that of most other coronaviruses. Various strains of MHV can infect the intestine, liver, spleen, respiratory tract, macrophages, and/or central nervous system (Virelizier *et al.*, 1975; Knobler *et al.*, 1981; Dubois-Dalcq *et al.*, 1982; Barthold and Smith, 1989). Classical studies by Bang and his coworkers showed that inbred strains of mice differ markedly in susceptibility to MHV, and that macrophages from resistant mouse strains are resistant to infection with MHV *in vitro* (Bang and Warwick, 1960). These observations have been extended, and three strains of mice with reduced susceptibility to MHV have been identified: C3H mice, in which only limited virus replication occurs and MHV infection is not fatal (Bang and Warwick, 1960; Arnheiter *et al.*, 1982; Woyciechowska *et al.*, 1984); A/J mice in which virus replication can be as extensive as in fully susceptible BALB/c mice, but with much less histopathology and disease (Arnheiter *et al.*, 1982; Dindzans *et al.*, 1986); and SJL/J mice, in which virus replication is profoundly reduced and MHV infection is not fatal (Smith *et al.*, 1984; Knobler *et al.*, 1985; Barthold and Smith, 1984).

We developed a solid phase virus binding assay to compare binding of the enterotropic and hepatotropic A59 strain of MHV to isolated membranes from MHV-susceptible BALB/c mice and MHV-resistant SJL/J mice (Boyle *et al.*, 1987). MHV-A59 virus bound to intestinal brush border membranes and hepatocyte membranes isolated from BALB/c mice, but not to comparable membrane preparations from SJL/J mice. Treatment of BALB/c membranes with neuraminidase or detergents such as NP40 or deoxycholate did not inhibit virus binding, but treatment of membranes with proteases did inhibit virus binding. When membrane proteins from MHV-susceptible BALB/c mice, MHV-resistant SJL/J mice and semi-resistant

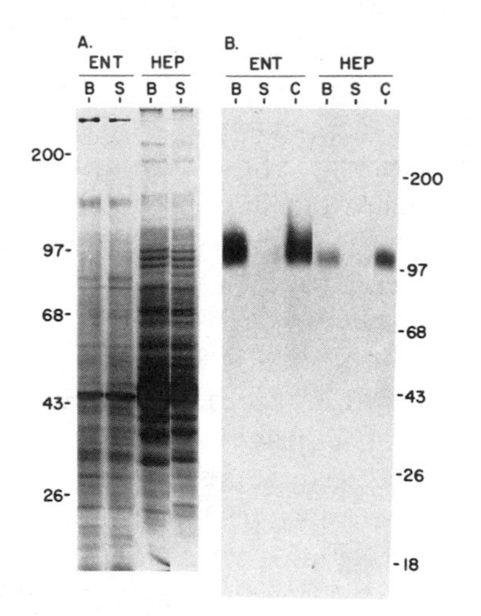

Figure 1. Virus-binding activity of membrane proteins from intestinal brush border membranes and hepatocyte membranes from strains of mice which differ in genetic susceptibility to mouse hepatitis virus (MHV). A. Coomassie blue staining of plasma membrane proteins from enterocytes (ENT) and hepatocytes (HEP) of BALB/c (lanes B) and SJL/J (lanes S) mice. B. Virus overlay protein blot assay of plasma membrane proteins from genetically susceptible BALB/c (lanes B), semi-resistant C3H (lanes C), and resistant SJL/J (lanes S) mice was performed by probing electroblotted membrane proteins with MHV-A59 and detecting bound virus with anti-viral antibody and redioiodinated staphylococcal protein A. Molecular weight standards (in kDa) are shown on right. Reproduced from Boyle *et al.*, 1987 with permission.

C3H mice were reduced, separated by SDS-polyacrylamide gel electrophoresis, and blotted to nitrocellulose, MHV-A59 virions bound to a single band of 100-120K from BALB/c and C3H intestine and liver (Figure 1). No virus-binding band was observed in SJL/J membranes (Boyle *et al.*, 1987). The virus-binding material from detergent-solubilized BALB/c intestine bound to several lectins (Holmes *et al.*, 1987). These data showed that MHV-A59 binds to a 100-120 kDa glycoprotein in the membranes of the natural target tissues of highly susceptible BALB/c mice and semi-resistant C3H mice, but that membranes of MHV-resistant SJL/J do not contain a virus-binding glycoprotein. Thus, the profound resistance of SJL/J mice to MHV may be due to absence of a functional receptor for the virus (Boyle *et al.*, 1987).

We developed monoclonal antibodies (MAb) directed against the MHV receptor glycoprotein (K. V. Holmes, in preparation). One of these, MAb CC1, was particularly interesting. It bound specifically to the 100 to 120 kDa glycoprotein in immunoblots of BALB/c liver and intestine membranes, and failed to recognize any glycoprotein in membranes from SJL/J mice (Figure 2; Williams *et al*, 1990). When mouse fibroblast cell lines were pretreated with MAb CC1 and then challenged with MHV-A59 virus, the cells were protected from infection (K. V. Holmes, in preparation). Binding of radiolabeled virus to cells was also blocked by pretreatment of cells with MAb CC1 (S. Snyder, in preparation). This anti-receptor MAb bound to mouse tissues with the same mouse strain specificity and tissue specificity observed in MHV-A59 virus infections. MAb CC1 showed strong immunofluorescent labeling of brush border membranes of BALB/c small intestine, and radiolabeled MAb CC1 bound to membranes of colon, small intestine and liver of BALB/c mice, but MAb CC1 failed to bind to membranes from SJL/J mice (R. K. Williams, 1991).

MAb CC1 was used for affinity purification of the MHV receptor glycoprotein from Swiss Webster mouse liver (Figure 2; Williams *et al.*, 1990). The isolated receptor was highly acidic, with a pI of approximately 4. It was also highly glycosylated, and treatment of the isolated receptor with peptide-N-glycanase F reduced the receptor to approximately 70 kDa (R. K. Williams, in preparation). N-terminal amino acid sequencing of the affinity purified MHV receptor was performed, and a synthetic peptide (NTR) corresponding to the first 15 amino acids of the receptor was synthesized. Anti-NTR antibody prepared in rabbits was used to immunoblot purified MHV receptor and intestinal and liver membranes from BALB/c and SJL/J mice (Figure 2B). Anti-NTR identified the 100 to 120 kDa MHV receptor in the BALB/c membranes. However, unlike MHV virus or MAb CC1, anti-NTR antibody identified a glycoprotein in the intestine and liver of SJL/J mice. In SDS-PAGE, this SJL/J glycoprotein migrated slightly faster than the MHV receptor glycoprotein of BALB/c mice. Binding of the anti-NTR antibody to BALB/c or SJL/J glycoproteins was specifically inhibited by absorption of the antibody with NTR peptide (Figure 2D). Thus, SJL/J mice

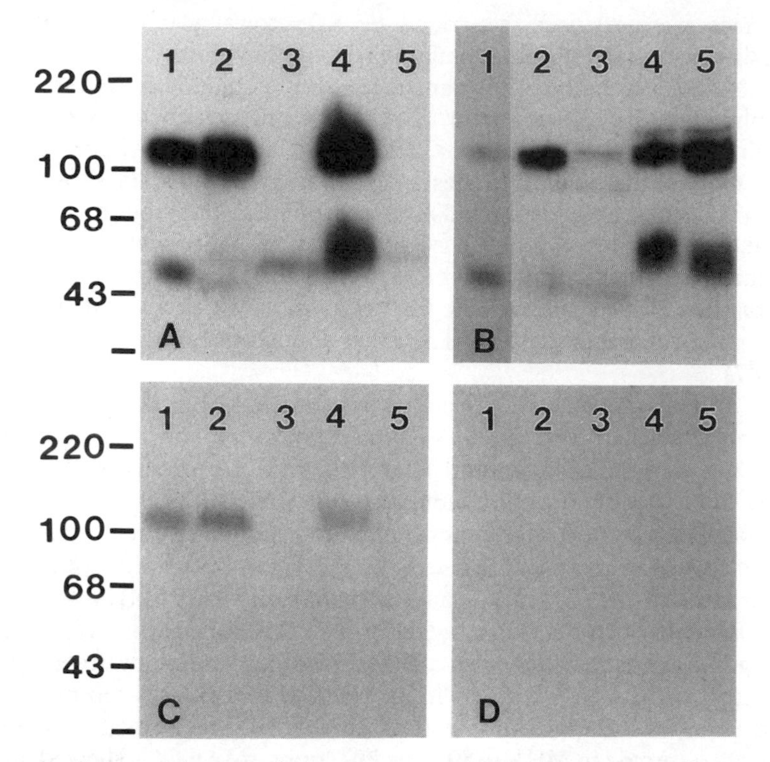

Figure 2. Immunoblot and virus overlay protein blot analysis of affinity-purified MHV receptor from Swiss Webster mouse liver and of BALB/c and SJL/J mouse liver and intestinal membranes. Panels A through D: Proteins were immunoblotted with anti-receptor MAb CC1 (A); antibody directed against the N-terminal peptide of the MHV receptor, anti-NTR (B); virus overlay protein blot, probed with MHV-A59 (C); or with anti-NTR that had been pre-adsorbed with 100ug of free NTR peptide per ml serum (D). Lanes: 1, 3ng of affinity purified membrane protein; 2 and 3, 100ug of liver membrane P2 fraction from BALB/c and SJL/J mice, respectively; 4 and 5, 100ug of intestinal brush border membranes from BALB/c and SJL/J mice, respectively. Reproduced from Williams *et al.*, 1990 with permission.

express a glycoprotein which shares at least the N-terminal epitope with the MHV receptor from BALB/c mice. This SJL/J molecule may be the homolog of the BALB/c glycoprotein which MHV recognizes as its receptor. The domain of the BALB/c glycoprotein which is recognized by MAb CC1 and MHV virus may be deleted or mutated in the SJL/J mouse so that the homologous glycoprotein fails to recognize these ligands. The normal cellular function of the MHV receptor glycoprotein from BALB/c mice is not yet known. Possibly the SJL/J homolog of the MHV receptor is able to perform its normal cellular function even though it cannot bind virus. This would be possible if the MHV-binding domain of the receptor glycoprotein is different from the domain required for its cellular function. Indeed, human immunodeficiency virus 1, rhinoviruses and Epstein Barr virus bind to their glycoprotein receptors at sites which are different from the functional domains of the glycoproteins (Jameson *et al.*, 1988; Clayton *et al.*, 1989; Lineberger *et al.*, 1990; Racaniello, 1990; Carel *et al.*, 1990; Barel *et al.*, 1990).

A glycoprotein of approximately 48 to 58 kDA which is immunologically cross-reactive with the MHV receptor glycoprotein was observed in immunoblots of intestine and liver membranes from both BALB/c and SJL/J mice (Figure 2; Williams *et al.*, 1990). The 48 to 58 kDa glycoprotein from BALB/c mice was recognized by anti-receptor MAb CC1 and anti-NTR antibody. It also bound MHV-A59 virions, but to a much lesser extent than the 110 kDa receptor. Intestinal brush border membranes from SJL/J mice had a homologous glycoprotein of about 50 kDa which was recognized by the anti-NTR antibody but not by anti-receptor MAb CC1 or by MHV-A59 virions. These results suggest that the 48-58 kDa glycoproteins may be derived from the same genes as the 100-120 kDa MHV receptor of BALB/c mice and its SJL/J homolog. We do not yet know whether the smaller glycoproteins are derived by proteolytic cleavage of the 100 to 120 kDa glycoproteins or whether they are products of alternative transcription of the genes for the larger proteins. The biological function(s) of the smaller glycoproteins are not yet known.

Solid phase assays for binding of MHV-A59 to intestinal brush border membranes from a variety of species, each of which serves as the natural host of a coronavirus, showed that MHV binds only to membranes from the intestine of MHV-susceptible or semi-resistant mice and not to intestinal brush border membranes from other species (Holmes *et al.*, 1989). Anti-receptor MAb CC1 showed the same binding specificity (S. R. Compton, in preparation). Thus, MHV binds to a species-specific domain of the 100-120 kDa glycoprotein receptor. We are now studying the receptors for coronaviruses of other species to determine whether they are related to the MHV receptor glycoprotein. These studies will show whether various coronaviruses have evolved to recognize species-specific determinants of a glycoprotein homologous to the 100-120 kDa MHV receptor or whether they recognize different types of receptors on membranes from different species.

Acknowledgements

We are grateful for the excellent technical assistance of C. Cardellichio, P. Elia, and S. Wetherell and for consultation of our colleagues Drs. S. Compton, C. Dieffenbach, M. Pensiero, M. Frana, and Ms. G. Dveksler and S. Gagneten. This work was supported in part by PHS grants # AI-18997, # AI-25231, and # AI-26075. The opinions expressed are the private views of the authors and should not be construed as official or necessarily reflecting the views of the Uniformed Services University of the Health Sciences or the Department of Defense.

References

Ang, F.B., and Warwick, A. (1960). Mouse macrophages as host cells for the mouse hepatitis virus and the genetic basis of their susceptibility. *Proc. Natl. Acad. Sci. U.S.A.* 46, 1065-1075.

Arnheiter, H., Baechi, T., and Haller, O. (1982). Adult mouse hepatocytes in primary monolayer culture express genetic resistance to mouse hepatitis virus type 3. *J. Immunol.* 129, 1275-1281.

Barel, M., Fiandino, A., Delcayre, A.X., Lyamani, F., and Frade, R. (1990). Monoclonal and anti-idiotypic anti-EBV/C3d receptor antibodies detect two binding sites, one for EBV and one for C3d on glycoprotein 140, the EBV/C3dR, expressed on human B lymphocytes. *J. Immunol.* 141, 1590-1595.

Barthold, S.W., and Smith, A.L. (1984). Mouse hepatitis virus strain-related patterns of tissue tropism in suckling mice. *Arch. Virol.* 81, 103-112.

Barthold, S.W., and Smith, A.L. (1989). Virus strain specificity of challenge immunity to coronavirus. *Arch. Virol.* 104, 187-196.

Barthold, S.W., Smith, A.L., Lord, P.F., Bhatt, P.N., Jacoby, R.O., and Main, A.J. (1982). Epizootic coronaviral typhlocolitis in suckling mice. *Lab. Anim. Sci.* 32, 376-383.

Barthold, S.W., Smith, A.L., and Povar, M.L. (1985). Enterotropic mouse hepatitis virus infection in nude mice. *Lab. Anim. Sci.* 35, 613-618.

Boyle, J.F., Weismiller, D.G., and Holmes, K.V. (1987). Genetic resistance to mouse hepatitis virus correlates with absence of virus-binding activity on target tissues. *J. Virol.* 61, 185-189.

Carel, J.C., Myones, B.L., Frazier, B., and Holers, V.M. (1990). Structural requirements for C3d,g/Epstein-Barr virus receptor (CR2/CD21) ligand binding, internalization, and viral infection. *J. Biol. Chem.* 265, 12293-12299.

Clayton, L.K., Sieh, M., Pious, D.A., and Reinherz, E.L. (1989). Identification of human CD4 residues affecting class II MHC versus HIV-1 gp120 binding. *Nature* 339, 548-551.

Dindzans, V.J., Skamene, E., and Levy, G.A. (1986). Susceptibility/resistance to mouse hepatitis virus strain 3 and macrophage procoagulant activity are genetically linked and controlled by two non-H2-linked genes. *J. Immunol.* 137, 2355-2360.

Dubois-dalcq, M.E., Doller, E.W., Haspel, M.V., and Holmes, K.V. (1982). Cell tropism and expression of mouse hepatitis viruses (MHV) in mouse spinal cord cultures. *Virology* 119, 317-331.

Holmes, K.V., Boyle, J.F., Weismiller, D.G., Compton, S.R., Williams, R.K., Stephensen, C.B. and Frana, M.F. (1987). Identification of a receptor for mouse hepatitis virus. *Adv. Exp. Med. Biol.* 218, 197-202.

Holmes, K.V., Williams, R.K., and Stephensen, C.B. (1989). Coronavirus receptors. In:" Concepts in Viral Pathogenesis III" (Notkins, A., and Oldstone, M.B.A., Eds.) pp. 106-113. Springer-Verlag, New York.

Jameson, B.A., Rao, P.E., Kong, L.I., Hahn, B.H. Shaw, G.M. Hood, L.E. and Kent, S.B. (1988). Location and chemical synthesis of a binding site for HIV-1 on the CD4 protein. *Science* 240, 1335-1339.

Knobler, R.L., Dubois-dalcq, M., Haspel, M.V., Claysmith, A.P., Lampert, P.W., and Oldstone, M.B. (1981). Selective localization of wild type and mutant mouse hepatitis virus (JHM strain) antigens in CNS tissue by fluorescence, light and electron microscopy. *J. Neuroimmunol.* 1, 81-92.

Knobler, R.L., Linthicum, D.S., and Cohn, M. (1985). Host genetic regulation of acute MHV-4 viral encephalomyelitis and acute experimental autoimmune encephalomyelitis in (BALB/cKe x SJL/J) recombinant-inbred mice. *J. Neuroimmunol.* 8, 15-28.

Lindsey, J.R. (1986). In:" Viral and Mycoplasmal Infections of Laboratory Rodents" (Bhatt, P.N., Jacoby, R.O., Morse, H.C.III, and New, A.E., Ed.) pp. 801. Academic Press, Inc., New York.

Lineberger, D.W., Graham, D.J., Tomassini, J.E., and Colonno, R.J. (1990). Antibodies that block rhinovirus attachment map to domain 1 of the major group receptor. *J. Virol.* 64, 2582-2587.

Racaniello, V.R. (1990). Cell receptors for picornaviruses. *Curr. Top. Microbiol. Immunol.* 161, 1-22.

Smith, M.S., Click, R.E., and Plagemann, P.G. (1984). Control of mouse hepatitis virus replication in macrophages by a recessive gene on chromosome 7. *J. Immunol.* 133, 428-432.

Spaan, W., Cavanagh, D., and Horzinek, M.C. (1988). Coronaviruses: Structure and genome expression. Review article. *J. Gen. Virol.* 69, 2939-2952.

Sturman, L.S., and Holmes, K.V. (1983). The molecular biology of coronaviruses. *Adv. Virus. Res.* 28, 35-112.

Virelizier, J.L., Dayan, A.D., and Allison, A.C. (1975). Neuropathological effects of persistent infection of mice by mouse hepatitis virus. *Infect. Immun.* 12, 1127-1140.

Wege, H., Siddell, S., and Ter Meulen, V. (1982). The biology and pathogenesis of coronaviruses. *Curr. Top. Microbiol. Immunol.* 99, 165-200.

Williams, R.K., Jiang, G.-S. and Holmes, K.H. (1991). Receptor for mouse hepatitis virus is a member of the carcino embryonic anatigen family of glycoproteins. *Proc. Natl. Acad. Sci. USA* 88, 5533-5536.

Williams, R.K., Jiang, G.-S., Snyder, S.W., Frana, M.F., and Holmes, K.V. (1990). Purification of the 110-kilodalton glycoprotein receptor for mouse hepatitis virus (MVH)-A59 from mouse liver and identification of a nonfunctional, homologous protein in MHV-resistant SJL/J mice. *J. Virol.* 64, 3817-3823.

Woyciechowska, J.L., Trapp, B.D., Patrick, D.H., Shekarchi, I.C., Leinikki, P.O., Sever, J.L. and Holmes, K.V. (1984). Acute and subacute demyelination induced by mouse hepatitis virus strain A59 in C3H mice. *J. Exp. Pathol.* 1, 295-306.

A Regulatory Cascade Controls Virulence in *Vibrio cholerae*

Victor J. DiRita, Claude Parsot, Georg Jander, and John J. Mekalanos

Introduction

Vibrio cholerae is the etiologic agent of the diarrheal disease cholera. In response to specific environmental conditions, this organism expresses several virulence determinants including the cholera toxin (Ctx), a toxin coregulated pilus (Tcp), and accessory colonization factor (Acf; DiRita and Mekalanos,1989; Taylor *et al.*, 1987; Peterson and Mekalanos, 1988). These gene products are part of a regulon under the control of the products of the toxRS locus of *V. cholerae* (Taylor *et al.*, 1987; Peterson and Mekalanos, 1988). ToxR is a transmembrane protein that binds specifically to the promoter for the operon encoding the cholera toxin (ctxAB) and activates transcription from this promoter in *Escherichia coli* (Miller and Mekalanos, 1985; Miller *et al.*, 1987).

Seventeen ToxR-activated genes (tag genes) have been defined by screening TnphoA fusions whose expression is coordinately regulated with that of cholera toxin (Peterson and Mekalanos, 1988). Most of the tag genes have been defined as being involved in Tcp and Acf biogenesis. In addition, the expression of two major outer membrane proteins, OmpU and OmpT, is controlled by ToxR (Miller and Mekalanos, 1988). Unlike ctxAB, the tag genes analysed so far are not activated by ToxR in *E. coli* (C. Parsot and J. Mekalanos, unpublished results) even though their expression is dependent on an intact toxR locus in *V. cholerae* (Peterson and Mekalanos, 1988). This has lead us to the conclusion that there may be other transcriptional activators responsible for the expression of genes in the ToxR regulon, and that these activators might themselves be under the control of ToxR (DiRita and Mekalanos, 1989). Such a "cascade" has been observed in sporulation by *Bacillus subtilis* where sigma factors that regulate endospore formation appear in sequential fashion as they are required for developmental gene expression (Losick and Pero, 1981). A regulatory cascade system like this is thought to be an efficient way to control expression of sets of genes during different stages of development.

In this chapter we summarize data on the control of toxR transcription by the heat shock response and the cloning of toxT, a new regulatory gene of *V. cholerae*. Like ToxR, the toxT gene product can activate the ctx promoter in *E. coli*. When expressed from a vector promoter, ToxT can direct coordinate expression of Ctx and Tcp in *V. cholerae* even in the absence of ToxR. Furthermore, several genes whose expression is dependent on ToxR in *V. cholerae* can be activated in *E. coli* by ToxT but not by ToxR. Consistent with the model of a regulatory cascade operating in *V. cholerae* virulence, we show that expression of toxT is controlled at the transcriptional level by ToxR.

Results

The htpG Gene is Located Upstream from toxR

The nucleotide sequence of the region upstream of the toxR gene has been determined for two strains of *V. cholerae* (569B and E7946). The sequences were found to be identical for the 600 bases pairs examined. Sequence analysis of the toxR upstream region indicated the presence of an open reading frame oriented opposite to toxR. This open reading frame displayed a high degree of similarity (approximately 86% identity) to the sequences of the heat shock proteins of the Hsp90 family. We have thus called this gene htpG, after the gene encoding the Hsp90 homolog in *E. coli* (Neidhardt and VanBogelen, 1987).

The htpG and toxR Promoter Are Regulated by Temperature

The promoter region upstream of the *V. cholerae* htpG gene is highly similar to the known consensus for *E. coli* heat shock promoters. This class of promoters is recognized by an alternate sigma factor (RpoH) which is elevated in level and activity by increased growth temperature (Neidhardt and VanBogelen, 1987). Accordingly, we tested the ability of the toxR and htpG promoters to be controlled by growth temperature. This was done by analyzing the expression of lacZ fusions to the toxR and htpG promoters over a range of temperatures. The conclusion of these studies was that toxR expression is reduced by elevated growth temperature while htpG expression responds in an opposite fashion.

The rpoH Gene Product Controls Expression htpG and toxR

We wished to test the hypothesis that the rpoH gene product was responsible for the temperature regulation we observed above. To do this, we utilized a plasmid construct (kindly provided by Dr. Carol Gross) that expresses RpoH under control of the tac promoter (pDS2). We found that in *E. coli*, the htpG and toxR promoters were also controlled by temperature in the same way as in *V. cholerae*. However, addition of the inducer IPTG to cultures of strains harboring pDS2 led to immediate derepression of the htpG promoter and simultaneous repression of the toxR promoter even at low growth temperature. We conclude that the close proximity of these two promoter has resulted in divergent promoters that are controlled simply by the level of the RpoH modified RNA polymerase in the cell. It follows that the level of the ToxR virulence regulator should be controlled by the heat shock response and indeed we have observed that all ToxR-regulated genes are reduced in expression at elevated temperatures consistent with this conclusion. Thus, control toxR transcription by heat shock represents the first level of cascade control affecting virulence properties of *V. cholerae*.

ToxR Mutants of *V. cholerae* Strain 569B Maintain Virulence Gene Expression

Expression of virulence factors such as Ctx and Tcp is modulated in most strains of *V. cholerae* in response to environmental conditions and is dependent on the product of the toxR gene (Taylor *et al*., 1987). The classical strain 569B, however, does not modulate Ctx expression to the same degree as other strains in response to changes in the growth environment although it does contain an active toxR gene. Strain 569B carries a deletion of the toxS gene, whose product enhances the activity of ToxR in both *E. coli* and *V. cholerae* (Miller *et al*., 1989). That 569B is able to express ToxR regulated genes without ToxS suggests that control of the ToxR regulon is altered in this strain relative to other strains of *V. cholerae*.

This was confirmed by comparing phenotypes of toxR mutants derived from strain 569B and O395. Introduction of the toxR55 insertion mutation into 569B had very little effect on expression of TcpA and CtxB. In contrast, introduction of the toxR55 mutation into strain O395 abolished expression of CtxB and TcpA Thus, strain 569B is apparently less dependent on ToxR for expression of Tcp and Ctx than strain O395. This result also suggests that strain 569B may express a regulator distinct from ToxR that is capable of activating ctx and tcp promoters.

Isolation of the toxT Gene from *V. cholerae* 569B

In order to identify activators other than ToxR that can activate the ctx promoter, we repeated the original screen used to clone toxR (Miller and Mekalanos, 1984). Briefly, plasmid DNA was isolated from a library of strain 569B and used to transform VM2, a ctx::lacZ reporter strain of *E. coli*, to ampicillin resistance. Transformants were scored for elevated expression of ctx::lacZ and the *V. cholerae* DNA inserts of these activating plasmids were compared with toxRS-encoding plasmids by restriction digestion and Southern blotting. We found that several of them bore similarities to each other but were different than plasmids harboring toxR and have thus termed the gene expressed by these plasmids toxT.

Activation of ToxR Regulated Genes by toxT

Because ToxT, like ToxR, could activate ctx expression in *E. coli*, we wished to determine whether ToxT would also activate other members of the ToxR regulon. The expression of three genes of the tcp cluster from strain O395, tcpA, tcpC, and tcpI (Taylor *et al.*, 1988) were found to be activated in *E. coli* by a plasmid encoding ToxT, but not by one encoding ToxR. Likewise, ToxT activated two other genes under ToxR control in *V. cholerae*, tagA (ToxR activated gene A) and aldA (the *V. cholerae* gene encoding aldehyde dehydrogenase; C. Parsot and J. Mekalanos, submitted for publication). The tagA gene was induced 70-fold and aldA 7-fold by ToxT, but neither were directly activated by ToxR alone. As expected, both ToxR and ToxT activated the ctx promoter in *E. coli*. Thus, while there is overlap in the ToxR regulon such that ToxR and ToxT can both activate ctx in *E. coli*, only ToxT directly activates several other genes (tcpA, tcpC, tcpI, aldA, and tagA) whose expression requires ToxR in *V. cholerae*.

Complementation of a toxR Mutant of *V. cholerae* by toxT

The dependence of tag gene expression on ToxR in *V. cholerae*, but on ToxT in *E. coli*, suggests that ToxR exerts control over tag expression through ToxT, perhaps by controlling toxT expression in *V. cholerae*. If this is so, then constitutive expression of toxT in a toxR mutant of *V. cholerae* might restore expression of ToxR regulated genes.

To test this, we took advantage of our observation that toxT expresses from plasmid pGJ2.3 is apparently under the control of the constitutive tet gene promoter from the cloning vecter. Plasmid pGJ2.3 was mobilized into wild-type V. choleraeO395 and its ToxR$^-$ deletion derivative JJM43. Strains O395, O395/pGJ2.3, JJM43, and JJM43/pGJ2.3 were grown under conditions that favor expression of the ToxR regulon (Miller and Mekalanos, 1988). As expected, strain JJM43 produced nearly undetectable levels of both Ctx and TcpA. While there was no detectable change in CtxB or TcpA expression in O395/pGJ2.3, introduction of pGJ2.3 into JJM43 restored CtxB and TcpA expression. This experiment suggests that the requirement of ToxR for ctx and tcp expression in *V. cholerae* can be eliminated by constitutive expression of a second regulatory gene, toxT.

Transcription of toxT Depends on ToxR

Given that transcription of toxT from a constitutive promoter overcomes a ToxR deficiency in *V. cholerae*, the most direct explanation for the observations presented in the previous section is that ToxR controls the expression of toxT whose product then activates several genes under indirect control by ToxR.

To investigate this possibility, we analyzed mRNA from O395(toxR$^+$) and JJM43(toxR$^-$) *V. cholerae*. As a probe, we used an EcoRV restriction fragment from pGJ2.3 into which mapped a majority of TnphoA insertions that abolish ToxT activity (V. DiRita and J. Mekalanos, unpublished). We found no detectable toxT mRNA in JJM43 but abundant toxT mRNA in O395. This indicates that transcription of the toxT gene or the stability of the toxT mRNA is under the control of ToxR.

Discussion

The toxR gene of *V. cholerae* was originally identified by its ability to directly activate expression of the ctx promoter in *E. coli* (Miller and Mekalanos, 1984). Subsequent studies have shown that toxR is required for expression of several other virulence genes in *V. cholerae* (Taylor *et al*., 1987; Peterson and Mekalanos, 1988), but direct control of their transcription by ToxR in a heterologous host such as *E. coli* has not been demonstrated. Identification of the toxT gene as described in this report may account for how ToxR activates expression of genes in *V. cholerae* it can not activate in *E. coli*.

Five genes originally identified as being under ToxR control in *V. cholerae* are, in *E. coli*, activated by ToxT but not by ToxR. ToxR-regulated gene products like Tcp and Ctx are not expressed in *V. cholerae* toxR mutants (Taylor *et al*., 1987). However, plasmid pGJ2.3, which constitutively expresses ToxT, can complement a *V. cholerae* toxR deletion mutant JJM43 for expression of Tcp and Ctx. Thus, while expression of genes in the ToxR regulon is dependent on ToxR under normal circumstances, these genes can be activated in the toxR$^-$ background by constitutively expressing ToxT. The dependence on ToxR in a wild-type *V. cholerae* background is evidently at the level of toxT transcription, as indicated by the absence of detectable toxT mRNA in the *V. cholerae* toxR mutant JJM43.

The data presented in this report are thus consistent with a model of regulatory cascading in which the heat shock sigma factor RpoH controls ToxR transcription and then ToxR activates expression of toxT. The toxT gene product then directly activates expression of several ToxR controlled virulence genes. The cascade presumably begins with negative control of toxR transcription by the heat shock response. We suggest this because the high temperature and adverse environmental conditions encountered in the human gastrointestinal tract such as exposure to low pH, anaerobiosis and starvation, would most likely shift *V. cholerae* into a heat shock state. This state would presumably be optimal for survival of this initial environment where expression of ToxR-activated gene are not needed or in fact may be deliterious. Eventually, *V. cholerae* would enter the small bowel and use motility and chemotaxis to move itself to a more favorable location (i.e., the intestinal epithelium) where activation of ToxR might occur in response to environmental signals even at low levels of toxR expression. Once this occurs, toxR would activate toxT and it in turn would activate expression of genes encoding cholera toxin and TCP pili. This last phase of ToxT-mediated expression might occur even in the face of new environmental signals that antagonize toxR transcription or activity (such as high temperature of host tissues).

The ToxR regulatory cascade may expand beyond the rpoH, toxR and toxT genes. For example, one of the genes shown to be activated by ToxT in this report is the tcpI gene, which itself appears to be a regulator of Tcp expression. If ToxT and other possible "downstream regulators" such as TcpI are themselves responsive to specific environmental conditions, then this system of cascading regulatory proteins would be a powerful mechanism for *V. cholerae* to fine tune expression of virulence determinants throughout its pathogenic cycle. Such a regulatory strategy might help virulent microorganisms deal with the rapid changes in environment that occur as a consequence of the pathophysiology of infection. Cascading may also be important to confer orderly temporal control over gene expression in much the same way as has been observed in developmental cycles such as sporulation.

References

DiRita, V.J. and Mekalanos, J.J., 1989, Genetic Regulation of Bacterial Virulence, *Ann. Rev. Genet.* 23: 455-482.

Losick, R. and Pero, J., 1981, Cascades of sigma factors. *Cell* 25:528-584.

Miller, V. L., DiRita, V. J., and Mekalanos, J. J., 1989, Identification of toxS, a regulatory gene whose product enhances ToxR-mediated activation of the cholera toxin promoter. *J. Bacteriol.* 171:1288-1293.

Miller, V. L., and Mekalanos, J. J., 1984, Synthesis of cholera toxin is positively regulated at the transcriptional level by toxR. *Proc. Natl. Acad. Sci. USA* 81:3471-3475.

Miller, V. L., and Mekalanos, J.J., 1988, A novel suicide vector and its use in construction of insertion mutations: osmoregulation of outer membrane proteins and virulence determinants in *Vibrio cholerae* requires toxR. *J. Bacteriol.* 170:2575-2583.

Miller, V. L., Taylor, R. K., and Mekalanos, J. J., 1987, Cholera toxin transcriptional activator ToxR is a transmembrane DNA binding protein. *Cell* 48:271-279.

Neidhardt, F.C. and VanBogelen, R.A., 1987, *Escherichia coli* and *Salmonella typhimurium*, In: Cellular and Molecular Biology, eds. Neidhart, F.C., Ingraham, J.L., Low, K.B., Magasanik, B., Schaecter, M. and Umbarker, H.E. (American Society for Microbiology, Washington, DC) pp. 1334-1345.

Peterson, K. M., and Mekalanos, J. J., 1988, Characterization of the *Vibrio cholerae* ToxR regulon: Identification of novel genes involved in intestinal colonization. *Infect. Immun.* 56: 2822-2829.

Taylor, R., Shaw, C., Peterson, K., Spears, P. and Mekalanos, J., 1988, Safe, live *Vibrio cholerae* vaccines? *Vaccine* 6: 151-154.

Taylor, R. K., Miller, V. L., Furlong, D. B., and Mekalanos, J.J., 1987, Use of phoA gene fusions to identify a pilus colonization factor coordinately regulated with cholera toxin. *Proc. Natl. Acad. Sci. USA* 84: 2833-2837.

Escape from the Phagosome and Cell-to-Cell Spread of *Listeria monocytogenes*

Daniel A. Portnoy, Andrew N. Sun, and Jacek Bielecki

Introduction

Listeria monocytogenes is a gram positive, rapidly growing, facultative, intracellular bacterial pathogen which occurs free-living in nature as well as in association with a variety of warm-blooded animals (Gellin *et al.*, 1989). The oral route is the natural route of transmission in humans as a number of well-documented cases have been traced to ingestion of contaminated foods. In fact, the presence of *L. monocytogenes* in food has become a major concern of the food industry.

Although the reported incidence of listeriosis is relatively low (approximately 2000 cases per year in the United States), there is a high case fatality rate in suseptible individuals including pregnant women, newborns, and immunocompromised individuals. In the latter the disease often presents as meningitis. It is surprising to some that the incidence among AIDS patients is relatively low (Jacobs *et al.*, 1985), although it is clearly more prevalent among AIDS patients than the general population (Gellin *et al.*, 1989).

Until recently, most of the interest in *L. monocytogenes* was due to it being an excellent model for the study of cell-mediated immunity in mice. In this model, an intravenous sublethal infection results in approximately 90% of the bacteria localizing in the liver within 30 min after injection. Most of these bacteria are killed during the next 6h, presumably in the Kupffer cells which line the sinusoids of the liver. This is followed by approximately two days of logarithmic bacterial growth followed by the rapid and complete elimination of the bacteria over the next eight to twelve days. (Newborg *et al.*, 1980). Since the pioneering work of Mackaness (Mackaness, 1962), the macrophage is usually considered to be the primary host cells for *L. monocytogenes* replication (Hahn *et al.*, 1978). However, the bulk of bacterial replication probably occurs in the parenchymal cells in the liver (North, 1970; Rosen *et al.*, 1989), and in intestinal epithelial cells during the normal route of infection (Racz *et al.*, 1972). Therefore, even though most models of anti-listerial immunity invoke macrophage activation, any model which deals with anti-listerial effector mechanisms must address the fact that the bacteria replicate in non-professional phagocytic cells. Also, as will be described below, the bacteria can spread from one host cell to another without leaving the cytoplasm.

Although there is a wealth of information on the host response to *L. monocytogenes* infection, until recently almost nothing was known about bacterial determinants of pathogenesis or the cell biology of intracellular growth. Two technical advances have led to a rapid increase in our understanding of the pathogenesis of *L. monocytogenes*. First, intracellular growth assays were developed in a number of cell-types including fibroblasts (Havell, 1986), the Caco-2 enterocyte-like cell line (Gaillard *et al.*, 1987), the Henle 407 epithelial cell line (Portnoy *et al.*, 1988) and the J774 macrophage-like cell line (Portnoy *et al.*, 1988). In summary, *L. monocytogenes* enters all of these cell-lines and grows with an intracellular doubling time of approximately 1 h which is close to the growth rate in rich medium. The second technical advance was the introduction of transposable elements into *L. monocytogenes* and the isolation of insertion mutants (Gaillard *et al.*, 1986; Kathariou *et al.*, 1987; Cossart *et al.*, 1989; Camilli *et al.*, 1990). Thus defined mutants can now be readily isolated and characterized with respect to pathogenicity in a mouse model and in a tissue culture model of infection.

Cell Biology of Intracellular Growth

It was first observed by Havell (1986) that *L. monocytogenes* invaded fibroblasts *in vitro* and spread cell-to-cell. Thus, in the presence of gentamicin and a soft agar overlay, an individual bacterium divided and apparently spread to the neighboring cells eventually forming a macroscopic plaque. The ability of *L. monocytogenes* to spread cell-to-cell is highly reminiscent of the intracellular growth behavior of *Shigella flexneri* in tissue culture which was also known to form plaques in monolayers of tissue culture cells (Oaks *et al.*, 1985). In addition, the spread of *S. flexneri* could be inhibited by cytochalasin D, a known inhibitor of actin polymerization (Pal *et al.*, 1989). Similarly, the cell-to-cell spread of *L. monocytogenes* could be inhibited by cytochalasin D at concentrations which had no measurable affect on bacterial internalization (Tilney *et al.*, 1989). A direct role for actin polymerization in the cell-to-cell spread of *S. flexneri* was shown by Bernardini *et al.* who showed that intracellular *S. flexneri* associated with polymerized actin in the cytoplasm of infected cells, and presented data which suggested that it was the association with actin which mediated the intra- and intercellular spread of the bacteria (Bernardini *et al.*, 1989).

The observations made on *S. flexneri* led investigators working on *L. monocytogenes* to investigate in detail the cell biology of *L. monocytogenes* growth and cell-to-cell spread in a macrophage cell-line using electron microscopy and video microscopy (see Fig. 1) (Tilney *et al.*, 1989; Dabiri *et al.*, 1990; Mounier *et al.*, 1990). The findings can be summarized as follows.

Subsequent to internalization, the bacteria escape from a host vacuole and are then found naked in the cell cytoplasm. Rapid cell division ensues, and the bacteria become encapsulated by short actin filaments and at least two other actin binding proteins, alpha actinin and tropomyosin. The actin filaments were identified as such using the techniques of electron microscopy and S1-decoration which uses the sub-fragment 1 of myosin to identify filaments as actin, and to show the direction of actin polymerization within a single filament (Tilney *et al.*, 1989). The actin based structure rearranges by an unknown mechanism to form a tail behind the bacteria which appears to mediate movement of the bacteria through the cytoplasm to the cell periphery. Intracellular bacteria can be measured moving at rates of approximately 1 micron per second, and the addition of cytochalasin D causes the rapid cessation of movement (Dabiri *et al.*, 1990). Next, the bacteria are presented in pseudopod-like structures which appear to shoot from the cell, and in some instances can grow to enormous lengths (Fig. 2). These bacterial-containing pseudopods are often recognized by a neighboring cell, and internalized. Thus, within the cytoplasm of the second cell, the bacteria can be found surrounded by a double membrane. Both membranes are dissolved, and the cycle is repeated.

One of the most satisfying consequences of the above model is that it provides a cell-biological explanation for the old observation that antibody plays little or no role in a murine model of infection. Thus, the bacteria may never need to leave the host cytoplasm to spread cell-to-cell. In fact, light microscopic examination of infected liver reveals a picture which appears quite similar to that seen *in vitro* (Robert North, personal communication). The proposed mechanism also emphasizes the importance of immunological effector mechanisms which can prevent the spread of infection and clear the bacteria from the cytoplasm of infected parenchymal cells, *i.e.* granuloma formation and CTL.

Role of Hemolysin in the Pathogenesis of Infection

Although there are likely to be numerous genes and gene-products necessary for the pathogenicity of *L. monocytogenes*, the hemolysin determinant, listeriolysin O, has been the most extensively studied. Transposon mutations resulting in loss of listeriolysin O secretion results in complete loss of virulence for mice, i.e., the LD_{50} increases by approximately 5 logs (Gailliard *et al.*, 1986, Kathariou *et al.*, 1987, Portnoy *et al.*, 1988). Introduction of the cloned gene encloding listeriolysin O into a hemolysin minus background (hly^-) fully restores virulence (Cossart *et al.*, 1989). Thus, listeriolysin O is clearly an essential determinant of pathogenicity.

Hly^-mutants exhibit an impaired growth in a variety of host cells (Gallard *et al.*, 1987; Kuhn *et al.*, 1988; Portnoy *et al.*, 1988). Electron microscopic analysis has revealed that the hly^- mutants are found in host vacuoles suggesting that the role of listeriolysin O is to lyse the phagocytic vacuole thus permitting access to the host cytoplasm (Gaillard *et al.*, 1987; Tilney *et al.*, 1989). Interestingly, the pH optimum for the lysis of erythrocytes by listeriolysin O is pH 5.6 which is consistent with its proposed site of action, the phagolysosome (Geoffroy *et al.*, 1987). However, weak bases such as chloroquine which raise the pH of acidic compartments (Maxfield, 1982) do not prevent *L. monocytogenes* from gaining access to the host cytoplasm (unpublished observations). Therefore, the importance of an acidic pH for the activity of hemolysin is unclear.

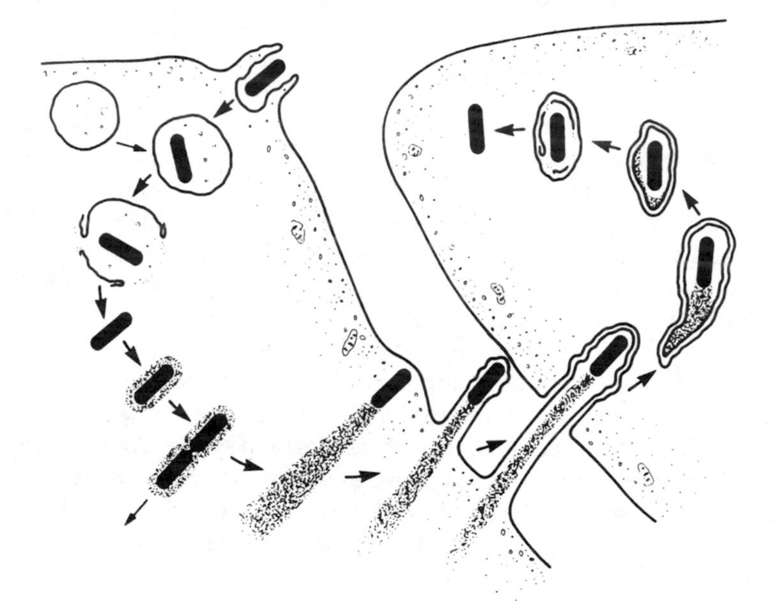

Figure 1. Morphological stages in the entry, growth, movement, and spread of *L. monocytogenes* from one cell to another. The fuzz represents actin filaments. (Tilney and Portnoy, 1989).

Although listeriolysin O is required for growth in most cells examined, there are exceptions. For example, hly⁻ mutants still grow in Henle 407 human epithelial cells (Portnoy *et al.*, 1988). This result suggests that other determinants can replace the activity of listeriolysin O in some cell lines. The significance of this observation is not yet appreciated.

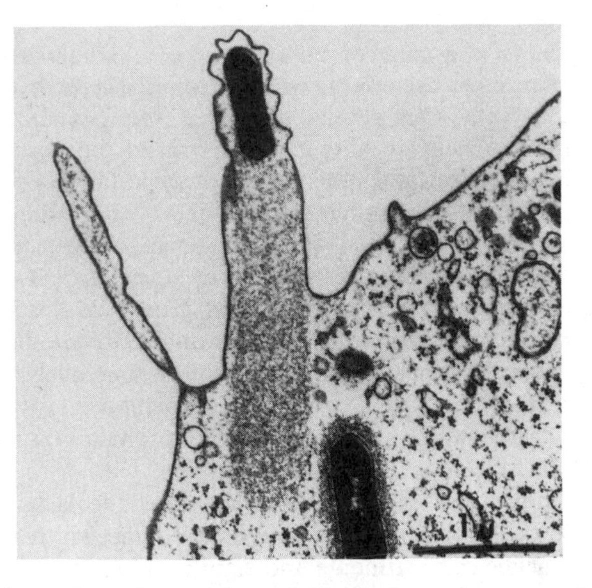

Figure 2. Thin section of a portion of the surface of a macrophage with *L. monocytogenes* being presented in a pseudopod-like structure. Notice the actin-based structure which trails the bacterium. (Tilney and Portnoy, 1989).

Expression of Listeriolysin O in *Bacillus subtilis*

To directly test the role of listeriolysin O, we cloned the structural gene, hlyA, into an asporogenic mutant of *Bacillus subtilis* where it was expressed under the control of an IPTG-inducible promoter referred to as the SPAC cassette (Bielecki *et al.*, 1990). The resulting strain, *B. subtilis* (hlyA), secreted detectable hemolytic activity only during growth in the presence of IPTG. These data indicated that a single gene product was sufficient for production and secretion of hemolytic activity by *B. subtilis*.

Next, we asked if hemolytic expression by *B. subtilis* would function inside cells to promote growth of the bacteria. The results (Fig. 3) showed that *B. subtilis* could grow in the J774 cells only when expressing listeriolysin O. In addition, electron microscopy revealed that *B. subtilis* expressing listeriolysin O was located mostly in the cytoplasm, while the control strain was localized entirely in membrane-bounded vacuoles. These data strongly support the model that the role of listeriolysin O is to lyse the phagocytic vacuole. In addition, these data indicate that the cytoplasm of J774 cells is a permissive environment for bacterial growth.

However, in subsequent experiments using primary cultures of bone marrow-derived macrophages, we were unable to demonstrate any bacterial replication. Thus, listeriolysin O is sufficient for escape from a J774 vacuole, but the normally non-pathogenic *B. subtilis*, even when expressing listeriolysin O, was unable to withstand the hostile intracellular enviroment of a primary macrophage.

Listeriolysin O is a member of a family of thiol-activated cytolysins found in gram positive bacteria in which streptolysin O is the prototype (Smyth *et al.*, 1978). Our results suggested the possibility that other members of this family could also mediate lysis of the host vacuole. To directly test this, the structural gene for streptolysin O (Kehoe *et al.*, 1987) was cloned into *B. subtilis* under control of the IPTG-inducible SPAC cassette as previously described (Bielecki *et al.*, 1990) to generate a streptolysin O producing strain referred to as *B. subtilis* (SLO). We have been unable to demonstrate any intracellular growth of this strain in J774 cells (Fig. 3) suggesting that streptolysin O is unable to lyse the phagosome. These results suggest that perhaps listeriolysin O is uniquely suited for the vacuolar environment. The most likely explanation for this is that listeriolysin O has a low pH optimum while streptolysin O is most active at neutral pH (Geoffroy *et al.*, 1987).

If low vacuolar pH was the explanation for the lack of activity of streptolysin O, then weak bases such as chloroquine which are known to concentrate in acidic compartments and significantly raise the vacuolar pH

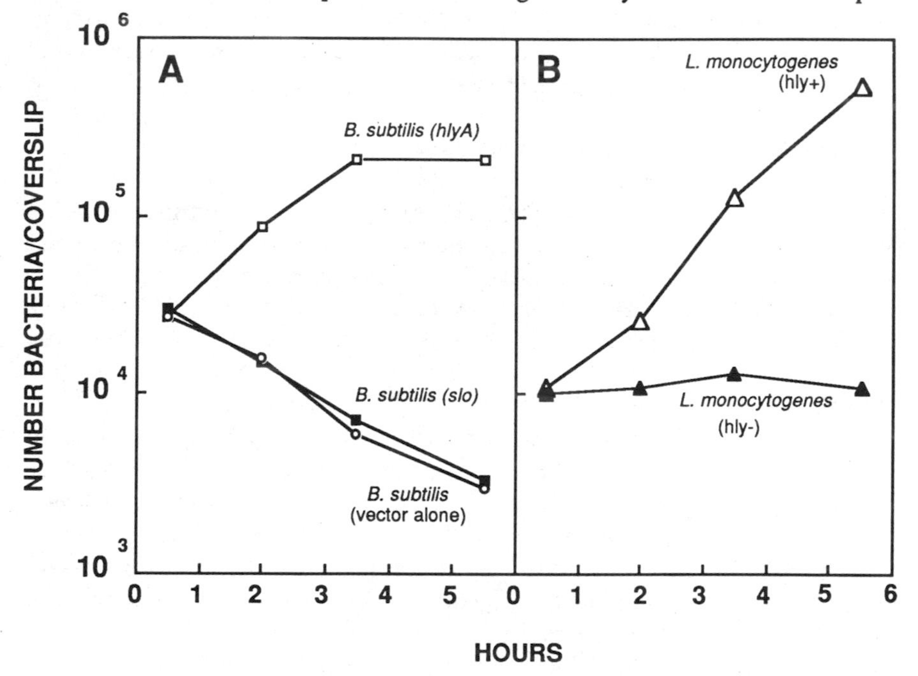

Fig. 3. Growth of *B. subtilis* strains expressing listeriolysin O or streptolysin O (A) and *L. monocytogenes* strains (B) in the J774 macrophage-like cell line. *B. subtilis* (hlyA) + IPTG, opened boxes; *B. subtilis* (SLO) + IPTG, closed boxes; *B. subtilis* with SPAC expression vector + IPTG, opened circles; wild type *L. monocytogenes*, opened triangles; hemolysin negative *L. monocytogenes*, closed triangles.

(Maxfield, 1982) might be expected to activate streptolysin O. Indeed, preliminary data suggest that *B. subtilis* (SLO) was able to grow in J774 cells when infected in the presence of chloroquine. This suggests that streptolysin O may be able to lyse the phagosome when the pH is raised.

Discussion

Intracellular parasites can be classified into those that reside within a host vacuole and those which grow directly in the host cytoplasm. *L. monocytogenes* is a member of the latter group which also includes most species of *Rickettsia*, *Shigellae* and *Trypanosoma cruzi* (Winkler *et al.*, 1988; Sansonetti *et al.*, 1986; and Ley *et al.*, 1990). Interestingly, all of these pathogens possess hemolytic activity which has been associated with the ability to enter the host cytoplasm. *Trypanosoma cruzi*, for example, secretes a pore-forming hemolysin which requires a low pH for activity. Consistent with this, weak bases which raise the pH of acidic compartments prevent access of T. cruzi to the host cytoplasm (Ley *et. al.*, 1990). Although listeriolysin O also has a pH optimum of approximately 5.6, weak bases do not prevent access of *L. monocytogenes* to the cytoplasm. In addition hemolysin mutants are able to grow in some cell lines. Therefore, it appears that although listeriolysin O is clearly an essential determinant of pathogenicity, which in some cells can mediate lysis of the vacuolar membrane, there are other factors produced by *L. monocytogenes* which in some cases can replace hemolysin.

B. subtilis expressing listeriolysin O is able to lyse the phagosome and grow rapidly in the host cytoplasm. However, although *B. subtilis* (hlyA) grew inside cells, there was no evidence for association with F-actin or cell-to-cell spread. Clearly, *L. monocytogenes* must possess determinants in addition to listeriolysin O which mediate cell-to-cell spread. In order to dissect the regulatory and structural requirements for cell-to-cell spread, we have designed a protocol based on transposon mutagenesis and the isolation of small-plaque mutants. (Sun *et al.*, 1990). These mutant fall into four general classes: 1) hly^- mutants unable to grow because they are defective in escape from the phagosome; 2) mutants which enter the cytoplasm, but which show abortive intracellular growth; 3) mutants which cannot associate with actin and hence cannot spread within a cell or cell-to-cell; and 4) mutants which spread within a cell, but are defective for cell-to-cell spread; Mutants in the last class have suffered transposon insertions in the structural gene encoding a phoshatidylinositol-specific phospholipase C (Camilli, Goldfine, and Portnoy, submitted). It seems quite possible that this phospholipase and/or another distinct phospholipase activity known to be secreted by *L. monocytogenes* may act in concert with listeriolysin O to mediate lysis of the vacuole and to mediate cell-to-cell spread.

References

Bernardini, M.L., *et al.*, 1989, icsA, a plasmid locus of *Shigella flexneri*, governs bacterial intra- and intercellular spread through interaction with F-actin, *Proc. Natl. Acad. Sci. USA*, 86:3867-3871.

Bielecki, J., *et al.*, 1990, Bacillus subtilis expressing a haemolysin gene from Listeria monocytogenes can grow in mammalian cells, *Nature* 345:175-176.

Camilli, A., Portnoy, D.A., and Youngman, P., 1990, Insertional mutagenesis of *Listeria monocytogenes* with a novel Tn917 derivative that allows direct cloning of DNA flanking transposon insertions, *J. Bacteriol.* 172:3738-3744.

Cossart, P., *et al.*, 1989, Listeriolysin O is essential for virulence of *Listeria monocytogenes*: direct evidence obtained by gene complementation, *Infect. Immun.* 57:3629-3636.

Dabiri, G.A., *et al.*, 1990, *Listeria monocytogenes* moves rapidly through the host cell cytoplasm by inducing directional actin assembly, *Proc. Natl. Acad. Sci. USA.* 87:6068-6072.

Gaillard, J.L., Berche, P., and Sansonetti, P., 1986, Transposon mutagenesis as a tool to study the role of hemolysin in the virulence of *Listeria monocytogenes*, *Infect. Immun.* 52:50-55.

Gaillard, J.L., *et al.*, 1987, In vitro model of penetration and intracellular growth of *Listeria monocytogenes* in the human enterocyte-like cell line Caco-2, *Infect. Immun.* 55:2822-2829.

Gellin, B. G., and Broome, C.V., 1989, Listeriosis, *J. Am. Med. Assoc.* 261:1313-1320.

Geoffroy, C., *et al.*, 1987, Purification, characterization, and toxicity of the sulfhydryl-activated hemolysin listeriolysin O from *Listeria monocytogenes*, *Infect. Immun.* 55:1641-1646.

Hahn, H., and Kaufman, S.H.E., 1981, The role of cell-mediated immunity in bacterial infections, *Rev. Infect. Dis.* 3:1221-1250.

Havell, E.A., 1986, Synthesis and secretion of interferon by murine fibroblasts in response to intracellular *Listeria monocytogenes*, *Infect. Immun.* 54:787-792.

Jacobs, J.L., and Murray, H.W., 1986, Why is *Listeria monocytogenes* not a pathogen in the acquired immunodeficiency syndrome? *Arch. Intern. Med.* 146:1299-1300.

Kathariou, S., *et al.*, 1987, Tn916-induced mutations in the hemolysin determinant affecting virulence of *Listeria monocytogenes*, *J. Bacteriol.* 169:1291-1297.

Kehoe, M.A., *et al.*, 1987, Nucleotide sequence of the streptolysin O (SLO) gene: structural homologies between SLO and other membrane damaging, thiol-activated toxins, *Infect. Immun.* 55:3228-3232.

Kuhn, M., Kathariou, S., and Goebel, W., 1988, Hemolysin supports survival but not entry of the intracellular bacterium *Listeria monocytogenes*, *Infect. Immun.* 56:79-82.

Ley, V., *et al.*, 1990, The exit of *Trypanosoma cruzi* from the phagosome is inhibited by raising the pH of acidic compartments, *J. Exp. Med.* 171:401-413.

Mackaness, G.B., 1962, Cellular resistance to infection, *J. Exp. Med.* 116:381-406.

Maxfield, F.R., 1982, Weak bases and ionophores rapidly and reversibly raise the pH of endocytic vesicles in cultured mouse fibroblasts, *J. Cell Biol.* 95:676-681.

Mounier, J., *et al.*, 1990, Intracellular and cell-to-cell spread of *Listeria monocytogenes* involves interaction with F-actin in the enterocyte-like cell line Caco-2, *Infect. Immun.* 58:1048-1058.

Newborg, M.F., and North, R.J., 1980, On the mechanism of T cell independent anti-listerial resistance in nude mice, *J. Immunol.* 124:571-576.

North, R.J., 1970, The relative importance of blood monocytes and fixed macrophages to the expression of cell-mediated immunity to infection. *J. Exp. Med.* 132:521-534.

Oaks, E.V., Wingfield, M.E., and Formal, S.B., 1985, Plaque formation by virulent *Shigella flexneri*, *Infect. Immun.* 48:124-129.

Pal, T., *et al.*, 1989, Intracellular spread of *Shigella flexneri* associated with the kcpA locus and a 140-kilodalton protein, *Infect. Immun.* 57:477-486.

Portnoy, D.A., Jacks, P.S., and Hinrichs, D.J., 1988, Role of hemolysin for the intracellular growth of *Listeria monocytogenes*, *J. Exp. Med.* 167:1459-1471.

Racz, P., Tenner, K., and Mero, E., 1972, Experimental *Listeria* enteritis. I. An electron microscopic study of the epithelial phase in experimental *Listeria* infection, *Lab. Invest.* 26:694-700.

Rosen, H., Gordon, S., and North, R.J., 1989, Exacerbation of murine listeriosis by a monoclonal antibody specific for the type 3 complement receptor of myelomonocytic cells. Absence of monocytes at infective foci allows *Listeria* to multiply in non-phagocytic cells, *J. Exp. Med.* 170:27-37.

Sansonetti, P.J., *et al.*, 1986, Multiplication of *Shigella flexneri* within HeLa cells: lysis of the phagocytic vacuole and plasmid-mediated contact hemolysis, *Infect. Immun.* 51:461-469.

Smyth, C.J., and Duncan, J. L., 1978, Thiol-activated (oxygen-labile) cytolysins. p. 129-183. In J. Jeljaszewicz and T. Wadstrom (ed.) Bacterial toxins and cell membranes. Academic Press, Inc. (London), Ltd., London.

Sun, A.N., Camilli, A., and Portnoy, D.A., 1990, Isolation of *Listeria monocytogenes* small-plaque mutants defective for intracellular growth and cell-to-cell spread, *Infect. Immun*. 58: 3770-3778.

Tilney, L. G., and Portnoy, D.A., 1989, Actin filaments and the growth movement, and spread of the intracellular bacterial parasite, *Listeria monocytogenes*, *J. Cell Biol.* 109:1597-1608.

Winkler, H.H., and Turco, J., 1988, *Rickettsia prowazekii* and the host cell: entry with, and control of the parasite, *Curr. Top. Microbiol. Immunol*. 138:81-107.

Global Regulatory Control of Curli Expression and Fibronectin Binding in Enterobacteria

Staffan Normark, Christina Ericson, Anna Jonsson, and Arne Olsen

Introduction

Many bacterial virulence-associated properties are affected by changes in the environment. For example, changes in growth temperature, osmolarity or oxygen tension may increase or decrease expression of adhesive organelles. Expression of *Vibrio cholerae* Tcp pili depends specifically on a transmembrane signal transmitter protein, ToxR (Miller *et al.*, 1987). Expression of other types of pili appears to be subjected to a more global regulatory control that may involve DNA topology. *Escherichia coli* Pap pili expression is thermoregulated such that these adhesive structures are formed at 37°C but not at room temperature (Goransson and Uhlin, 1984). This low temperature repression operates at the level of transcription and can be relieved by a *drd*X mutation in the structural gene for the histone-like protein, H1 (Goransson *et al.*, 1990). Mutations in this gene have been isolated in many laboratories using different screening or selection protocols (Higgins *et al.*, 1990), and the H1 gene has been given many designations, including *bgl*Y, *drd*X, *hns*, *pil*G, *osm*Z, and *vir*R. Mutations in H1 affect gene expression of osmoregulated genes such as *pro*U, hence the name *osm*Z for the locus (Higgins *et al.*, 1988). The *E. coli* K12 *bgl* operon for β-glucoside metabolism is normally cryptic, but can be activated by *bgl*Y mutations (Defez and DeFelice, 1981). The frequency of phase variation of *E. coli* type 1 pili is controlled by a DNA element that can be inverted during a site-specific recombination event. A *pil*G mutation increases the frequency of this switching event (Spears *et al.*, 1986). *Shigella flexneri* is normally noninvasive at room temperature, whereas strains carrying a *vir*R mutant allele are invasive (Dorman *et al.*, 1990). Higgins and colleagues have demonstrated that H1 mutants have altered DNA supercoiling and suggest that the different phenotypes associated with defects in H1 result from changes in DNA topology (Hulton *et al.*, 1990).

When grown at 26°C on media containing low NaCl concentration many *E. coli* isolates express thin, coiled fiber (Olsen *et al.*, 1989). Curliated *E. coli* binds soluble fibronectin. The laboratory strain HB101 is normally noncurliated and fails to bind fibronectin. When expressing the Crl protein from a recombinant plasmid, HB101 cells produce curli and bind soluble fibronectin in a thermoregulated manner. *E. coli* HB101 has no *crl* gene due to its *pro*A2 deletion, but other *E. coli* isolates and laboratory strains carry the *crl* gene whether or not they express curli. Initially, it was believed that the *crl* gene encoded curlin, the subunit protein of curli. However, the fact that the *crl* gene product lacks a cleavable signal peptide and has a primary sequence suggestive of a cytoplasmic location raises the possibility that the 15.7 kD Crl protein is a regulatory protein. If so, HB101 must carry the genes in a nonexpressed form for curli formation and for fibronectin binding which can be activated by the plasmid-borne *crl* gene.

Results and Discussion

The *E. coli crl* gene encodes a polypeptide 133 amino acid in size. The gene was sequenced from *E. coli* isolates AO12 and K12 (Olsen *et al.*, 1989). Only one amino acid difference was found, suggesting that this protein is not undergoing frequent sequence variation among *E. coli* strains. The *pho*E genes of *Salmonella typhimurium* and *Enterobacter cloacae* have been cloned (Van de Leg *et al.*, 1987; J. Tomassen, personal communication). As in *E. coli*, the *crl* gene from these species is located adjacent to *pho*E. The deduced amino acid sequences of the *crl* gene products from *E. coli*, *S. typhimurium*, and *E. cloacae* are highly homologous (more than 70% identity). Expression of recombinant *S. typhimurium* and *E. cloacae* Crl protein from plasmids in *E. coli* HB101 results in curli formation and soluble fibronectin binding activity.

Transcriptional studies have shown that the *E. coli crl* gene is part of a monocistronic operon (A. Jonsson, unpublished data). Curli are preferentially expressed at low temperature and in conditions of relatively low osmolarity. However, transcriptional activity of the *crl* gene, as monitored by Northern blot analysis and expression from a *crl-lac*Z transcriptional fusion, is not affected by these environmental parameters. Moreover, curliated and noncurliated *E. coli* show similar transcription of the *crl* gene. Finally, introducing *crl* on a multicopy plasmid into a curli-deficient *E. coli* strain does not lead to curliation even though the Crl protein is detectable by SDS-PAGE. These data suggest that expression of curli is not controlled by the level of Crl protein in the cell.

The *crl* gene lacks a signal peptide cleavage site and the amino terminus of the protein does not have the characteristics of a leader sequence, suggesting that Crl may not be a secretory protein. Immunoelectronmicroscopy provides further evidence to suggest that Crl is located primarily in the cytoplasm.

A series of stop codons was introduced into the *crl* gene, resulting in a series of truncated Crl polypeptides. In a truncated mutant missing the carboxy-terminal eighteen amino acids (115 residues in size), curli are formed. In another mutant composed of only 94 amino acids, curli formation is abolished. Thus, the carboxy terminus of Crl is not essential for biologic activity. This contrasts sharply with the pilus subunit proteins in which the carboxy-terminal sequence is essential for interaction with a periplasmic chaperone protein during pilus biogenesis (Hultgren *et al.*, 1989; Lindberg *et al.*, 1989).

All of these data suggest that cytoplasmic Crl is able to activate expression of cryptic curli genes in HB101. This is reminiscent of the cryptic *bgl* operon in *E. coli* K12 that can be activated by mutations in *bgl*Y, the structural gene for histone-like protein, H1 (Lejeune and Danchin, 1990). To test this similarity, a *bgl*Y mutant allele was introduced into a *rec*A$^+$ derivative of HB101. This H1 mutant (unlike the wild type parent producing intact H1) expresses large quantities of curli at 26°C but not at 37°C and is able to bind soluble fibronectin when growing at low temperature and low osmolarity. When the same *bgl*Y allele was introduced into other noncurliated *E. coli* K12 strains (such as MC1029) the mutant became curli-proficient and able to bind fibronectin. A *bgl*Y, *crl*$^-$ double mutant of MC1029 still expresses curli and binds fibronectin, but does so less well than the MC1029 *bgl*Y single mutant.

Curli expression in natural *E. coli* isolates is not confined to specific clones, since phylogenetically different strains express curli. The possibility therefore exists that curli expression is under a phase switch. We do not yet have clear evidence for such a switch in *E. coli*, but have compelling evidence for phase variation of curli in *Salmonella*. *Salmonella enteritidis* strain 27655-3a forms smooth colonies on CFA agar. A rough variant occurred spontaneously that did not have any alteration in the LPS side chains. While the smooth variant did not express curli and did not bind soluble fibronectin, the rough variant did. Colony morphology backswitching from rough to smooth occurred at a high frequency and resulted in loss of expression of curli and ability to bind soluble fibronectin.

In conclusion, both *E. coli* and *Salmonella* have the genetic capacity to express curli. Curli expression can be activated in the noncurliated strain HB101 by mutations in the H1 gene or by expression of the Crl protein. Irrespective of the means of activation, expression remains temperature-controlled. It is not known why many natural isolates of *E. coli* are curliated. It seems unlikely they should be natural H1 mutants. It may well be that expression of other regulatory proteins differs between curliated and noncurliated *E. coli*. It is also not clear whether curli are fibronectin-binding organelles or if a fibronectin-binding protein has the same kind of global control as does the expression of curli.

References

Defez, R., and DeFelice, M., 1981, Cryptic operon for β-glucoside metabolism in *Escherichia coli* K12: genetic evidence for a regulatory protein, *Genetics* 97: 11-25.

Dorman, C.J., NiBhriain, N., and Higgins, C.F. , 1990, DNA supercoiling and environmental regulation of virulence gene expression in *Shigella flexneri*, *Nature* 344: 789-792.

Goransson, M., Sonden, B., Nilsson, P., Dagberg, B., Forsman, K., Emanuelsson, K., and Uhlin, B.E., 1990, Transcriptional silencing and thermoregulation of gene expression on *Escherichia coli*, *Nature* 344: 682-685.

Goransson, M. and Uhlin, B.E., 1984, Environmental temperature regulates transcription of a virulence pili operon in *E. coli*, *EMBO J.* 3: 2885-2888.

Higgins, C.F., Dorman, C.J., Stirling, D.A., Waddell, L., Booth, I.R., May, G., and Bremer, E., 1988, A physiological role for DNA supercoiling in the osmotic regulation of gene expression in *S. typhimurium* and *E. coli*, *Cell* 52: 569-584.

Higgins, C.F., Hinton, J.C.D., Hulton, C.S.J., Owen-Hughes, T., Pavitt, G.D., and Seirafi, A., 1990, Protein H1: a role for chromatin structure in the regulation of bacterial gene expression and virulence?, *Mol. Microbiol.*, 4: 2007-2012.

Hultgren, S., Lindberg, F., Magnusson, G., Kihlberg, J., Tennent, J., and Normark, S., 1989, The PapG adhesin of uropathogenic *Escherichia coli* contains separate regions for receptor binding and for the incorporation into the pilus, *Proc. Natl. Acad. Sci.* 86: 4357-4361.

Hulton, C., Seirafi, A., Hinton, J., Sidebotham, J., Waddell, L., Pavitt, G., Owen-Hughes, T., Spassky, A., Buc, H., and Higgins, C., 1990, Histone-like protein H1 (H-NS) DNA supercoiling and gene expression in bacteria, *Cell* 63: 631-642.

Lejeune, P., and Danchin, A., 1990, Mutations in the *bgl*Y gene increase the frequency of spontaneous deletions in *E. coli*, *Proc. Natl. Acad. Sci. USA* 87: 360-363.

Lindberg, F., Tennent, J., Hultgren, S., Lund, B., and Normark, S., 1989, PapD a periplasmic transport protein in P-pilus biogenesis, *J. Bacteriol.* 171: 6052-6058.

Miller, V.L., Taylor, R. K., and Mekalanos, J., 1987, Cholera toxin transcriptional activator ToxR is a transmembrane DNA binding protein, *Cell* 48: 271-279.

Olsen, A., Jonsson, A., and Normark, S., 1989, Fibronectin binding mediated by a novel class of surface organelles on *Escherichia coli*, *Nature* 338: 652-655.

Spears, P.A., Schauer, D. and Orndorff, P.E., 1986, Metastable regulation of type 1 piliation in *Escherichia coli* and isolation and characterization of a phenotypically stable mutant, *J. Bacteriol.* 168: 179-185.

Van De Leg, P., Bekkers, A., Van Meersbergen, J., and Tomassen, J., 1987, A comparative study on the *pho*E genes of three enterobacterial species. Implications for structure-function relationships in a pore-forming protein of the outer membrane, *Eur. J. Biochem.* 164: 469-475.

Collagen Receptor of *Staphylococcus aureus*

Lech M. Switalski, Wade G. Butcher, Joseph M. Patti, Pietro Speziale, Anthony G. Gristina, and Magnus Hook

Introduction

Development of new antimicrobial agents and improved public health conditions have not substantially reduced the frequency of infections caused by *Staphylococcus aureus.* Staphylococci are the leading cause of osteomyelitis, septic arthritis, endocarditis, wound and foreign body infections. While, as with many other bacteria, increasing antibiotic resistance of *S. aureus* strains is notable, it is not the major cause of failure in the treatment of staphylococcal infections. What constitutes the challenge is that staphylococci may resist treatment with conventional antimicrobial agents because of difficulties in achieving and maintaining therapeutically active concentrations of antibiotics in tissues which are either poorly vascularized (like bone) or harbor colonies of bacteria where diffusion is impaired (such as biofilms on surgical implants). Paradoxically, advances in the field of medical therapy have contributed to a steady increase of in the number of infections caused by coagulase-positive and -negative staphylococci by creating more situations in which these opportunistic pathogens find a suitable environment to multiply.

First described by Ogston more than a century ago, *S. aureus* has remained a mysterious pathogen. It is obvious that its virulence is multifactorial, and related to the production of a wide variety of extracellular and cell surface bound pathogenicity factors. The relative importance of these factors may vary depending on the site and stage of an infection, and various models have been proposed to explain the roles of individual factors. The most effective way to prevent an infection is early in the process of infection, before bacteria manage to multiply.

The concept of preventing infection by interfering with the initial microbial adherence to the host tissue is in this context particularly appealing. The molecular mechanisms of microbial adherence have been extensively studied for many Gram-negative bacteria, and only in the last decade shed some light on possible adhesion mechanisms of Gram-positive bacteria. Historicaly speaking, the association of *S. aureus* with fibrinogen (mediated by what was subseqently termed clumping factor, or less correctly, bound coagulase) described by Much in 1908 appears to be the first description of a putative adherence mechanism.

When Kuusela (1978) described binding of *S. aureus* cells to the then newly rediscovered fibronectin, a new chapter in the study of *S. aureus* adherence was opened. Soon, many other observations followed, indicating that i. fibronectin may be recognized by many different bacteria, both Gram-positive and Gram-negative, and ii. in addition to fibronectin, many other connective tissue proteins are recognized and bound by bacteria. Currently the rate of description of new interactions between bacteria and connective tissue components, or eukaryotic cell surface components like integrins, seems to be limited only by the rate at which these components are discovered.

In addition to fibrinogen and fibronectin, *S. aureus* strains associate with several other adhesive eukaryotic proteins (many of which belong to the family of adhesive matrix proteins) - laminin (Lopes *et al.*, 1985), vitronectin (Chhatwal *et al.* 1987), bone sialoprotein (Ryden *et al.*, 1989), proteoglycans (Ryden *et al.*, 1989), endothelial cell membrane protein (Tompkins *et al.*, 1990) and collagens. These interactions have mostly been studied in systems in which either soluble host proteins or microparticles coated with these proteins are incubated with bacteria. Indications that these bacteria-protein interactions play a role in virulence have been demonstrated in only few instances. Progress in this field has been relatively slow and may require a detailed knowledge of the bacterial components that serve as receptors for the matrix proteins. Along this line the fibronectin receptor has been purified and the corresponding gene cloned and sequenced (Froman *et al.*, 1987; Signas *et al.*, 1989). The evaluation of the role of the fibronectin receptor as a virulence factor has been more difficult, although observations *in vitro* as well as indirect clinical data strongly suggest its importance in staphylococcal colonization and invasiveness (Proctor, 1987). In more complex *in vitro* systems, interference with fibronectin binding to staphylococci seems to significantly, but not completely, inhibit attachment of bacteria to various substrates (Raja *et al.*, 1990). The potential biological role of the *S. aureus* fibronectin receptor was recently evaluated in an *in vivo* endocarditis model. It was shown that a transposon mutant of *S. aureus* lacking the fibronectin receptor exhibited at least two orders of magnitude lower adherence to traumatized heart valves compared to the parent strain (Kuypers *et al.*,1989).

Collagens as Major Components of Cartilage and Bone

Cartilage and bone tissues have a fairly low density of cells and a high content of extracellular matrix primarily composed of collagens and proteoglycans. Collagens are the major structural proteins of vertebrates (for reviews see: Nimni, 1980; Bornstein *et al.*, 1980; von der Mark, 1981). They are generally defined as possessing a triple helical domain containing peptide chains with repeating Gly-X-Y triplets, and also containing hydroxyproline and hydroxylysine residues. A number of different collagen types have been discovered, all of which are clearly related but differ genetically, chemically and immunologically.

The most abundant collagens are type I and II collagens. They both form fibers, and arise from a precursor converted to collagen by the proteolytic removal of amino and carboxy terminal propeptides. Type I collagen is the major structural component of skin, bone, tendon and other fibrous tissues, where it forms large, banded fibers. Type II collagen is the major structural protein in cartilage. In contrast to the fiber forming collagens of type I and II, type IV collagen, specific for the basement membranes, forms an open, non-fibrillar network created through the association of like ends of the molecule. Additional minor components of bone and cartilage include, among others, collagens of other types and some tissue specific glycoproteins, such as osteonectin, osteopontin and 148 kD cartilage matrix glycoprotein (Heinegard *et al.*, 1988 and 1989). In tissues these components form an insoluble extracellular matrix. Interspersed in this matrix are the cells and, in the case of bone, inorganic depositions (Zambrano *et al.*, 1982).

Binding of Collagen to Cells of *Staphylococcus aureus*. Unique Features of the Collagen Binding Component

Staphylococcus aureus has been shown to be the organism that most frequently causes bacterial arthritis both in children (one third of cases) (Argen *et al.*, 1966; Nelson *et al.*, 1966; Jackson *et al.*, 1982) and adults (Argen *et al.*, 1966). *S. aureus* is also the dominating pathogen (half the cases) in the etiology of hematogenous (primary) and secondary osteomyelitis (Waldvogel *et al.*,1980; Petersen *et al.*, 1980; Emslie *et al.*, 1986). In infectious arthritis and osteomyelitis, bacterial cells adhere to and colonize synovial and bone tissues, respectively (Gristina *et al.*, 1985; Speers *et al.*, 1985).

Binding of staphylococci to collagen was first reported from several laboratories in the mid nineteen eighties (Vercellotti *et al.*, 1985; Holderbaum *et al.*, 1985; Carrett *et al.*, 1985). Subsequently, the binding of collagen to staphylococci has been characterized in some detail (Speziale *et al.*, 1986, Holderbaum *et al.*, 1986).

Staphylococci binding one collagen type also bind other types, which indicates that they recognize a common, repetitive sequence in the collagen molecule. In fact, binding of type II ^{125}I-collagen to bacteria can be inhibited not only by intact collagen, alpha chains and its large fragments, but also by polypeptides mimicking the repetitive motif of collagen ($[PGP]_n$, $[PPG]_n$,and $[PHyPG]_n$). The precise structure of the fragment(s) bacteria recognize has not yet been determined. The binding is highly specific, and unrelated proteins do not significantly inhibit the amount of collagen bound to the cells. A recent report on the inhibition of collagen binding to *S. aureus* by fucose (Buxton *et al.*, 1990) may indicate that bacteria also recognize carbohydrate moieties on collagen.

The number of measurable collagen binding sites varies from one strain to another, but is remarkably high, reaching 3 x 10^4 per cell in strain Cowan. This value is close to two orders of magnitude higher compared to that of the staphylococcal receptors recognizing fibronectin or fibrinogen (Hawiger *et al.*, 1982; Froman *et al.*, 1987). On the other hand, the affinity of binding appears to be lower, and an estimated K_d value of type II collagen binding to cells of *S. aureus* strain Cowan 1 is in the order of 10^{-7}M. The lower values of dissociation constants reported by others may be related to different strains of bacteria or methods of preparation (Holderbaum *et al.*, 1986; Buxton *et al.*, 1990).

Another striking feature of the collagen binding component is its immunogenicity. Antibodies raised against whole cells of staphylococci effectively inhibit binding of ^{125}I-collagen to bacteria. These antibodies are cross-reactive among different strains of S. *aureus*.

Isolation and Purification of *S. aureus* Collagen Receptor

The low affinity of binding is the main reason that affinity chromatography could not be used in the course of purification of the native receptor. Moreover, the low affinity solubilized receptor can not be assayed using standard competition assays. We have therefore developed a two step assay to quantify the solublized receptor. In this assay the test bacteria (*S. aureus* strain Cowan) are incubated with polyclonal antibodies raised against whole homologous bacterial cells. (Antibodies are absorbed with cells of *S. aureus* Newman - a strain which does not bind collagen). This step results in an almost complete blocking of collagen binding sites by the antibodies. If the incubation mixture also contains solubilized collagen receptor, the antibody inhibition can be neutralized, and within certain range of receptor concentrations, the reversal of the inhibition caused by antibodies is linear.

A three step purification procedure has been developed to purify the collagen binding component, and its activity at different steps has been measured using the assay described above. The purified product, isolated from *S. aureus* strain Cowan is a protein with M_r of 135 kD.

In agreement with our previous observations on the activities of antibodies raised against whole cells of collagen binding bacteria, polyclonal antibodies raised against the purified receptor inhibited the binding of ^{125}I collagen to bacteria. These monovalent anti-receptor antibodies recognized only one protein, in the lysostaphin lysate of *S. aureus* strain Cowan (Fig.1, lane b) out of several proteins present in the lysate (Fig.1, lane a). The band visualized with antibodies is identical to the band identified as the purified receptor. Moreover, these antibodies were cross-reactive, and detected only one major band on Western blots of whole cell lysostaphin digests of several strains of receptor positive strains of staphylococci (Fig. 1, lanes c through i),

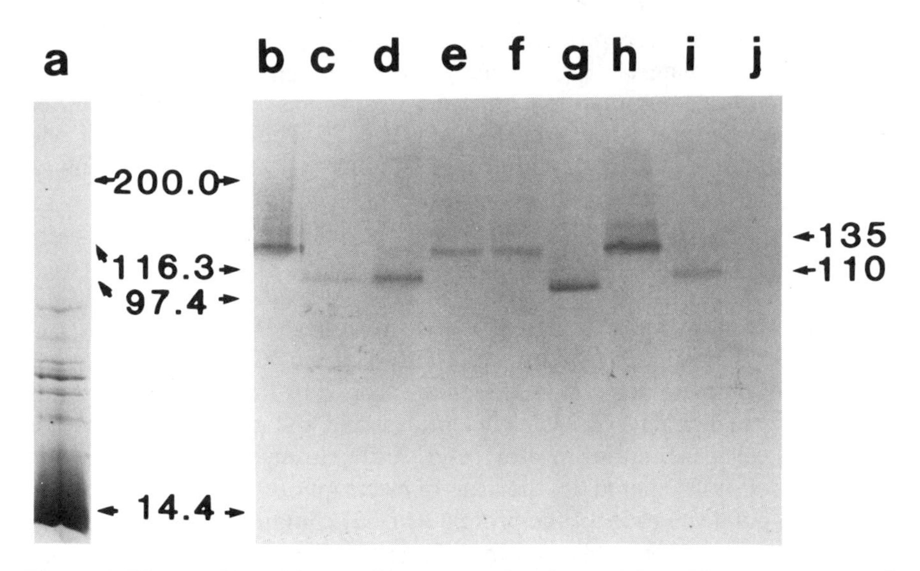

Figure 1. Western blot of lysostaphin lysates of various strains of *S. aureus* separated by electrophoresis and probed with anti-collagen receptor antibodies. Lysostaphin lysates were separated by gel electrophoresis in reducing conditions in 5-10% polyacrylamide gel. The separated material was electroblotted onto Immobilon P membrane (Millipore, Bedford, MA) and probed with anti-*S. aureus* strain Cowan collagen receptor monovalent polyclonal antibodies, followed by peroxidase conjugated secondary antibodies (Switalski *et al.*, 1989). Lane a: lysate of *S. aureus* Cowan, unblotted gel, stained with Coomassie brilliant blue G, lanes b through j blotted and probed with antibodies. Lysates of various strains of *S. aureus*- b: Cowan, c: 87/12 (infectious arthritis), d: 87/8 (infectious arthritis), e: X50151 (infectious arthritis), f: 87/2 (infectious arthritis), g: 88/4 (osteomyelitis), h: M65237 (osteomyelitis), i: Phillips (infectious arthritis). Lane j contains lysate of a strain M65051 (soft tissue infection), representative of nine receptor negative strains tested.

while lysates of receptor negative strains did not contain any immunoreactive material (Fig. 1, lane j). The apparent molecular weight of the immunoreactive component in some strains was 135 kD, while other strains expressed receptor with M_r of 110 kD. The differences in the receptor size do not seem to affect the binding capacity of the strains of origin.

Structure of the Collagen Receptor

We have recently cloned and sequenced the collagen adhesin gene *(cna)* from a λGT11 genomic library from *S. aureus* strain FDA 574 (Patti *et al.* 1992). The complete nucleotide sequence encoded by the cna gene contains an open reading frame of 3555 nucleotides, and the predicted M_r of the translated collagen binding protein of 133 kD closely correlates with the value previously calculated on the basis of electrophoretic mobility.

A putative model of the protein (Fig. 2) contains a signal sequence at the amino terminus, followed by a large nonrepetitive A domain and three highly homologous 187 amino acid long repeats (B). Immediately following the repeats is a 64 amino acid lysine and proline rich region. At the carboxy end, the receptor contains a long stretch of hydrophobic residues (M) followed by a few positively charged amino acids (C).

Although no significant amino acid homology with any known protein has been detected, the general structure of the collagen receptor resembles that of other surface proteins of *Staphylococcus aureus* including

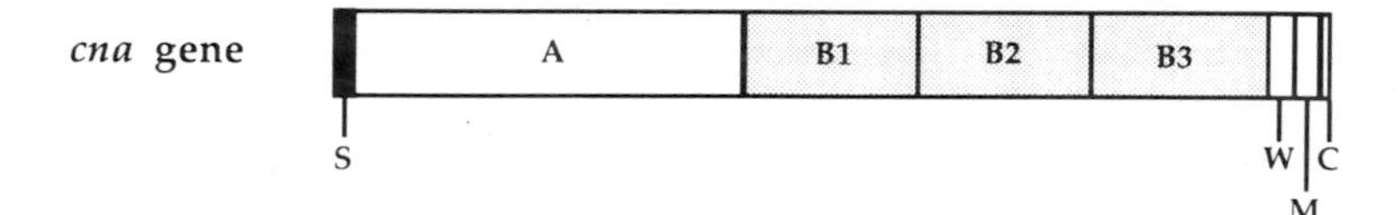

Figure 2. Postulated model of the collagen receptor from *S. aureus* FDA 574 (M_r 133 kD). S- signal sequence, A- large domain of unspecified function, B- three repeat domains, W- cell wall domain, M- membrane spanning domain, C- charged carboxy terminal domain.

protein A (Uhlen *et al.*, 1984) and the fibronectin binding protein (Signas *et al.*, 1989), or *Streptococcus pyogenes* - protein G (Guss *et al.*, 1986) and protein M (Hollingshead *et al.*, 1986). It is also noteworthy that the pentapeptide LKPTG motif, typical for the cell wall spanning domain of known surface proteins of Gram-positive bacteria (see: Fischetti *et al.* this volume) is also present in the collagen receptor, although in slightly modified form - LPTGM.

The Collagen Binding Protein as an Adhesin

Earlier observations indicated a remarkable correlation between the ability of bacteria to bind collagen and their ability to cause infection within collagen rich tissues like bone and cartilage. As observed by us and others, the great majority of *S. aureus* strains isolated from infections within bone (osteomyelitis) or synovial space (infectious or septic arthritis) bind collagen, while strains from other infections express collagen binding components with much lower frequency (Holderbaum *et al.*, 1987; Switalski *et al.*, 1991). These preliminary observations suggested that the presence of the collagen receptor determined to a large extent the ability of bacteria to cause these particular types of infections.

To evaluate the role of the collagen receptor in bacterial adherence we tested the ability of ^{125}I-labeled cells of various *S. aureus* strains to adhere to surfaces coated with type II collagen. Only strains able to bind soluble collagen adhered to these surfaces. Our data indicated that the process of attachment is essentially completed within 3 h. Receptor negative strains do not attach to collagen coated surfaces (Fig. 3 A, C). As expected anti-receptor antibodies almost completely abolished bacterial attachment to these surfaces (Fig. 3 B).

It should be mentioned that synthetic collagen-like peptides which inhibit the binding of ^{125}I-collagen to bacteria (see above) had only marginal effect on the attachment of cells to collagen coated surfaces. Although the reasons for this are unclear, it is possible that the multivalency of the collagen coated surface results in a dramatic increase in apparent affinity compared to the situation when an isolated peptide constitutes the ligand.

In another model, more closely resembling the early events in the development of infectious arthritis, we have evaluated bacterial attachment to cartilage. Earlier electron microscopy data of the specimens collected from experimentally infected joints indicated that the primary attachment site for bacteria in the joint is not the synovial membrane, but rather the cartilage (Voytek *et al.*, 1988). Similar data arose from the electron microscopy observations of attaching staphylococci in the course of experimental osteomyelitis (Speers *et al.*, 1985). The asay model which we have developed involves measuring of bacterial attachment to uniform pieces of bovine nasal cartilage.

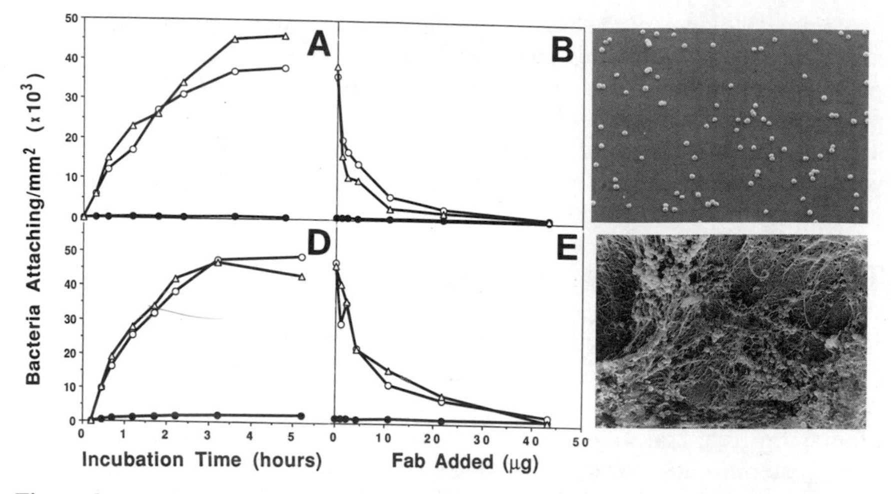

Figure 3. Attachment of selected strains of *S. aureus* to type II collagen coated surfaces (panels A, B, and C) or cartilage (panels D,E, and F). Symbols of *S. aureus* strains: Cowan, Phillips (both collagen receptor positive), M65051 - receptor negative.

Attachment to collagen coated surfaces. Detachable microtiter wells (Immulon 2 from Dynatech, Chantilly, Va) were incubated overnight at room temperature with type II collagen from chicken sternum (10 ug/well). Unsaturated binding sites were blocked with bovine serum albumin and the wells were overlayed with 100 ul of a suspension of ^{125}I labeled bacteria (2 x 10^7 cells). The wells were incubated with bacteria for the indicated periods of time (panel A) or for 2 h at room temperature. After extensive washing with phosphate buffer saline containing 0.1% Tween 80 the wells were individually counted in the gamma counter.

Attachment to cartilage. Uniform pieces of bovine nasal cartilage were incubated with 2 x 10^7 cells of ^{125}I labeled bacteria for the indicated periods of time (panel D) or for 2 h at room temperature. Cartilage pieces were washed with phosphate buffered saline containing 0.1% Tween 80 and individually counted in the gamma counter.

Inhibition of attachment by antibodies (panels B and E): F(ab) fragments of anti-*S. aureus* strain Cowan receptor antibodies (Switalski *et al.*, 1989) were incubated with bacteria for 1 h at 37°C prior to the incubation in wells or with cartilage.

Scanning electron microscopy (panels C and F) represents the attachment of *S. aureus* strain Cowan). Bars in both panels denote 10 um.

Adhesion data obtained with pieces of cartilage as the substrate basically followed those of collagen coated surfaces - in terms of kinetics, the number of attaching cells (Fig.3 D and F) and inhibition by antibodies (Fig. 3 E). Scanning electron microscopy of bacteria attaching to cartilage reveals that staphylococci preferentially attach to collagen fibers (Fig. 3 F). These data indicate that even in more complex substrates like cartilage which contain in addition to collagen a number of other components of unknown reactivity toward staphylococci, tissue adherence appears to involve a binding to collagen. This conclusion was supported by the observation that strains which do not express the receptor did not attach in appreciable numbers to the cartilage.

The Collagen Receptor is Sufficient for Mediating Attachment to Cartilage

Polystyrene beads similar in size to staphylococci (approx. 1 um diameter) were covalently coated with purified collagen receptor. These beads were subsequently used in the series of experiments to assess the importance of the collagen receptor in the attachment to cartilage.

Collagen receptor coated beads were able to bind ^{125}I-collagen (Fig. 4 A), but not unrelated ^{125}I-labeled proteins such as fibrinogen or fibronectin (data nor shown). Control beads coated with the same molar amount of a recombinant form of the staphylococcal fibronectin receptor did not bind ^{125}I-collagen. The kinetics of ^{125}I-collagen binding to collagen receptor coated beads resembled that of bacteria and binding was essentially complete within 30 min of incubation. This binding was blocked in the presence of anti-receptor antibodies (Fig. 4 B).

In attachment assays, both to collagen coated surfaces and cartilage, beads behaved very much as bacteria (Fig. 4C). The attachment was completely inhibited by antibodies against the collagen receptor (Fig. 4 D). While additional evidence may be needed to fully assess the importance of the collagen adhesin, particularly in *in vivo* experiments, these data strongly suggest that the interaction between the 110/135 kD receptor and cartilage collagens has an important biological role.

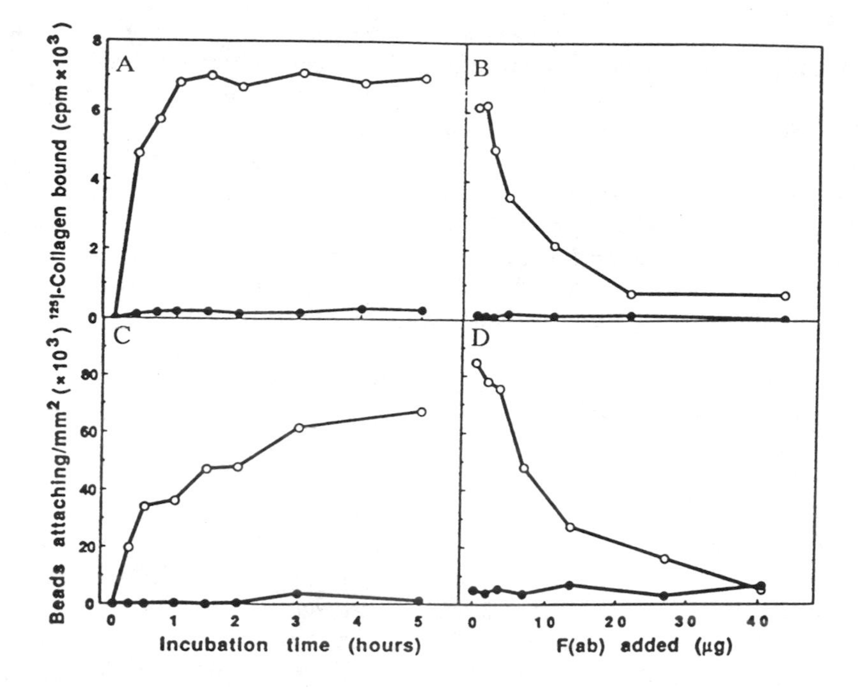

Figure 4. Binding of ^{125}I-collagen (panels A and B) or attachment to cartilage (panels C and D) of polystyrene beads coated with the collagen receptor of *S. aureus* strain Cowan (O) or the recombinant form of *S. aureus* fibronectin receptor (●). Data shown refer to the experiment in which each bead had approx. 2 x 10^4 receptor molecules on the surface. Panels A and C - time course of binding or attachment, panels B and D - inhibition of attachment by antibodies.

Binding of ^{125}I type II collagen to the beads followed the procedure described for bacterial cells (Speziale *et al.*, 1986), other procedures were as described in the legend to Fig. 3.

Perspectives

Recognition, and subsequent attachment of bacteria to collagen containing substrates may not be the only mechanism of bacterial attachment to host tissues. While the majority of *S. aureus* strains isolated from joint and bone infections bind collagen, and most likely use this interaction as a mechanism of tissue adherence, some strains do not bind collagen, and their adherence must depend on other mechanisms, *e.g.* recognition of bone sialoprotein or proteoglycans as it was recently postulated by Ryden *et al.* (1989).

The ability to bind collagen is not limited to strains of *S. aureus*. It also occurs among many strains of coagulase negative staphylococci, where it seems to some extent correlate with strain pathogenicity (Mamo *et al.*, 1988; Paulsson *et al.*, 1990; Watts *et al.* 1990) although other laboratories, using different assay procedures, do not confirm these observations (Holderbaum *et al.*, 1987). Several other species of bacteria also bind collagen. Binding of collagen by some strains of *E.coli* and salmonellae may be a mechanism of attachment of these bacteria to host tissue after damage to the intestinal epithelium and be a component ot the tissue invasion process (Gonzalez *et al.*, 1988; 1990; Visai *et al.*, 1990). The O75X adhesin of uropathogenic strains of *E. coli* has been shown to bind type IV collagen, and this phenomenon appears to be clearly correlated with the type of infections caused by these bacteria (Westerlund *et al.*, 1989). Oral bacteria implicated in the etiology of periodontal disease - *Bacteroides gingivalis* bind and attach to collagen (Naito *et al.*, 1988). The evaluation of this phenomenon may be difficult, since these organisms are strongly collagenolytic, and both binding and degradation of the substrate may take place concomitantly. Attachment of streptococci to dentin collagen was recently postulated to contribute to dental caries (Liu *et al.*, 1990 a and b) and recognition of type IV collagen in the glomerular basement membrane by *Streptococcus pyogenes* strains (Kostrzynska *et al.*, 1989) may be a step leading to the development of poststreptococcal glomerulonephrits. Collagen type II also binds to strains *Streptococcus pyogenes* (Speziale *et al.*, 1986; Mamo *et al.*, 1987) and yersiniae (Emody *et al.*, 1989) where an arthritic type of response sometimes follows infection. Clearly, the interactions of microorganisms with collagen will be explored in the future.

Acknowledgements

This project was supported by grants AI20624 from the National Institutes of Health and the Arthritis Foundation Biomedical Research Grant.

References

Argen, R.J., Wilson Jr., C.H., and Wood, P., 1966, Suppurative arthritis. Clinical features of 42 cases, *Arch. Intern.* Med.117:661-666.

Bornstein, P., and Sage, H.,1980, Structurally distinct collagen types, *Annu. Rev. Biochem.* 49:957-1003.

Buxton, T.B., *et al.*, 1990, Binding of *Staphylococcus aureus* bone pathogen to type I collagen, *Microbial Pathogenesis* 8:441-448.

Carret, G., *et al.*, 1985, Gelatin and collagen binding to *Staphylococcus aureus* strains, *Ann. Inst. Pasteur (Paris)* 136A:241-245.

Chhatwal, G.S., et al, 1987, Specific binding of the human S protein (vitronectin) to streptococci, *Staphylococcus aureus*, and *Escherichia coli*, *Infect. Immun.* 55:1878-1883.

Emody, L., *et al.*, 1989, Binding of collagen by *Yersinia enterocolitica* and *Yersinia pseudotuberculosis:* Evidence for yopA-mediated and chromosomally encoded mechanisms, *J. Bacteriol.* 171:6674-6679.

Emslie, K.R., and Nade, S., 1986, Pathogenesis and treatment of acute hematogenous osteomyelitis: evaluation of current views with reference to an animal model, *Revs. Infect. Dis.* 8:841-849.

Froman, G., *et al.*, 1987, Isolation and characterization of a fibronectin receptor from *Staphylococcus aureus*, *J. Biol. Chem.* 262:6564-6571.

Gonzalez, E.A., *et al.*, 1988, Growth conditions for the expression of fibronectin and collagen binding to *Salmonella*, *Zbl. Bakt. Hyg. A* 269:437-446.

Gonzalez, E.A., *et al.*, 1990, Virulent *Escherichia coli* strains for chicks bind fibronectin and type II collagen, *Microbios* 62:113-127.

Gristina, A.G., *et al.*, 1985, Adherent bacterial colonization in the pathogenesis of osteomyelitis, *Science* 228:990-993.

Guss, B., *et al.*, 1986, Structure of the IgG binding-regions of streptococcal protein G, *EMBO J.* 5:1567-1575.

Hawiger, J., *et al.*, 1982, Identification of a region of human fibrinogen interacting with staphylococcal clumping factor, *Biochemistry* 21:1407-1413

Heinegard, D., *et al.*, 1988, Noncollagenous matrix constituents of cartilage, *Pathol. Immunopathol. Res.* 7:27-31.

Heinegard, D., and Oldberg, A., 1989, Structure and biology of cartilage and bone matrix noncollagenous macromolecules, *FASEB J.* 32042-2051.

Holderbaum, D., Hall, G.S., and Ehrhart, L.A., 1986, Collagen binding to *Staphylococcus aureus*, *Infect. Immun.* 54:359-364.

Holderbaum, D., Spech, R.A., and Ehrhart, L.A., 1985, Specific binding of collagen to *Staphylococcus aureus*, *Collagen Rel. Res.* 5:261-271.

Holderbaum, D., *et al.*, 1987, Collagen binding in clinical isolates of *Staphylococcus aureus*, *J. Clin. Microbiol.* 25:2258-2261.

Hollingshead, S.K., Fischetti, V.A., and Scott, J.R., 1986, Complete nucleotide sequence of type 6 M protein of group A streptococcus. *J. Biol. Chem.* 261:1677-1686.

Jackson, M.A., and Nelson, J.D.,1982, Etiology and medical management of acute suppurative bone and joint infections in pediatric patients, *J. Pediatr. Orthoped.* 2:313-323.

Kostrzynska, M., Schalen, C., and Wadstrom, T., 1989, Specific binding of collagen type IV to *Streptococcus pyogenes*, FEMS Microbiology Letters 59:229-234.

Kuusela, P., 1978, Fibronectin binds to *Staphylococcus aureus*, *Nature (Lond.)* 276:718-720.

Kuypers, J.M., and Proctor, R.A., 1989, Reduced adherence to traumatized rat heart valves by a low-fibronectin-binding mutant of *Staphylococcus aureus*, *Infect. Immun.* 57:2306-2312.

Liu, T., and Gibbons, R.J., 1990, Binding of streptococci of the mutans group to type I collagen associated with apatitic surfaces, *Oral Microbiol. Immunol.* 5:131-136.

Liu, T., Gibbons, R.J., and Hay, D.I., 1990, *Streptococcus cricetus* and *Streptococcus rattus* bind to different segments of collagen molecules, *Oral Microbiol. Immunol.* 5:143-148.

Lopes, J.D., dos Reis, M., and Brentani, R.R., 1985, Presence of laminin receptors in *Staphylococcus aureus*, *Science* 229:275-277.

Mamo, W., Froman, G., and Wadstrom, T., 1988, Interaction of sub-epithelial connactive tissue components with *Staphylococcus aureus* and coagulase-negative staphylococci isolated from bovine mastitis, *J. Clin. Microbiol.* 27:540-544.

Mamo, W., *et al.*, 1987, Binding of fibronectin, fibrinogen and type II collagen to streptococci isolated from bovine mastitis, *Microbial Pathogenesis* 2:417-424.

Naito, Y., and Gibbons, R.J., 1988, Attachment of *Bacteroides gingivalis* to collagenous substrata, *J. Dent. Res.* 67:1075-1080.

Nelson, J.D., and Koontz, W.C., 1966, Septic arthritis in infants and children: a review of 117 cases, *Pediatrics* 38:966-971.

Nimni, M.E.,1980, The molecular organization of collagen and its role in determining the biophysical properties of the connective tissues, *Biorheology* 17:51-82.

Patti, J.M., Jonsson, H., Guss, B., Switalski, L.M., Wiberg, K, Lindberg, and Hook, M. (1992) Molecular characteriztion and expression of a gene coding for fibronectin-binding proteins from *Streptococcus dysgalactiae* and identification of active sites, *J. Bio. Chem.* In press.

Paulsson, M., and Wadstrom, T., 1990, Vitronectin and type-I collagen binding by *Staphylococcus aureus* and coagulase-negative staphylococci, *FEMS Microbiol. Immunol.* 65:55-62.

Petersen, S., *et al.*, 1980, Acute haematogenous osteomyelitis and septic arthritis in childhood. A 10-year review and follow up, *Acta Orthop. Scand.* 51:451-457.

Proctor, R.A., 1987, The staphylococcal fibronectin receptor: evidence for its importance, *Rev. Infect. Dis.* 9 (Suppl.4):335-340.

Raja, R., Raucci, G., and Hook, M., 1990, Peptide analogs fo a fibronectin receptor inhibit attachment of *Staphylococcus aureus* to fibronectin-containing substrates, *Infect. Immun.* 58:2593-2598.

Ryden, C., *et al.*, 1989, Specific binding of bone sialoprotein to *Staphylococcus aureus* isolated from patients with osteomyelitis, *Eur. J. Biochem.* 184:331-336.

Signas, C., *et al.*, 1989, Nucleotide sequence of the gene for a fibronectin-binding protein from *Staphylococcus aureus*: use of this peptide sequence in the synthesis of biologically active peptides, *Proc. Natl. Acad. Sci. USA* 86:699-703.

Speers, D.J., and Nade S.M.L., 1985, Ultrastructural studies of adherence of Staphylococcus aureus in experimental acute hematogenous osteomyelitis, *Infect. Immun.* 49:443-446.

Speziale, P., *et al.*, 1985, Binding of type II collagen to streptococci, *Zbl. Bakt. Hyg.* A

Speziale, P., *et al.*, 1986, Binding of collagen to *Staphylococcus aureus* Cowan 1, *J. Bacteriol.* 167:77-81.

Switalski, L.M., Patti, J.M., Butcher, W., Gristina, A.G., Speziale, P., and Hook, M., 1992, A collagen receptor on septic arthritis *Staphylococcus aureus* isolates mediates adhesion to cartilage, *Cell*, Submitted.

Switalski, L.M., Speziale, P., and Hook, M., 1989, Isolation and characterization of a putative collagen receptor from *Staphylococcus aureus* strain Cowan 1, *J. Biol. Chem.* 264: 21080-21086.

Tompkins, D.C., *et al.*, 1990, A human endothelial cell membrane protein that binds *Staphylococcus aureus in vitro*, *J. Clin. Invest.* 85:1323-1327.

Uhlen, M., *et al.*, 1984, Complete sequence of the staphylococcal gene encoding protein A, *J. Biol. Chem.* 259:1695-1702.

Vercellotti, G.M., *et al.*, 1985, Extracellular matrix proteins (fibronectin, laminin, and type IV collagen) bind and aggregate bacteria, *Amer. J. Pathol.* 120:13-21.

Visai, L., Speziale, P., and Bozzini, S., 1990, Binding of collagens to an enterotoxigenic strain of *Escherichia coli*, *Infect. Immun.* 58:449-255.

von der Mark, K., 1981, Localization of collagen types in tissues, *Review Connect. Tissue Res.* 9:265-324.

Voytek, A., *et al.*, 1988, Staphylococcal adhesion to collagen in intra-articular sepsis, *Biomaterials* 9:107-110.

Waldvogel, F.A., and Vasey, H., 1980, Osteomyelitis: the past decade, *New England J. Med.* 303:360-369.

Watts, J.L., Naidu, A.S., and Wadstrom, T., 1990, Collagen binding, elastase production, and slime production associated with coagulase-negative staphylococci isolated from bovine intramammary infections, *J. Clin. Microbiol.* 28:580-583.

Westerlund, B., *et al.*, 1989, The O75X adhesin of uropathogenic *Escherichia coli* is a type IV collagen-binding protein, *Molec. Microbiol.* 3:329-337.

Zambrano, N.Z., *et al.*, 1982, Collagen arrangement in cartilages, *Acta Anat.* 113:26-38.

Prospects for Group A Streptococcal Vaccine

Edwin H. Beachey and Malak Kotb

Introduction

Although the attack rate of acute rheumatic fever has declined dramatically in the industrialized nations of the world, it remains a rampant disease in third world countries and continues to be the major cause of heart disease in children around the world. Indeed, the prevalence rates of rheumatic heart disease in school-aged children is as high as 33 per thousand in the urban slums of some developing countries (Kholy *et al.*, 1978). A number of programmes have been undertaken for the prevention and control of rheumatic fever and rheumatic heart disease in various regions around the world. The major emphasis of these programmes has been on secondary prevention, including long-term antibiotic prophylaxis to prevent recurrent attacks of acute rheumatic fever. Primary prevention programmes directed toward prevention of the initial attack of rheumatic fever also have been instituted, but with less success. These have emphasized early diagnosis of streptococcal infections with prompt antibiotic treatment of the acute infections. Inasmuch as the diagnosis of streptococcal infections is often difficult to make in many parts of the world, and mass antibiotic prophylaxis is impractical if not impossible, the primary prevention programmes have been far from successful. It is clear that additional measures are required to prevent attacks of rheumatic fever that can be triggered by asymptomatic streptococcal infections or by infections for which medical care is not sought. Toward this end, efforts have been made for many years by a number of investigators to develop a safe and effective vaccine against strains of group A streptococci that give rise to acute rheumatic fever and rheumatic heart disease. These earlier vaccination efforts have been thwarted by toxic reactions to almost any streptococcal product administered to human volunteers (Stollerman, 1975). Furthermore, some streptococcal vaccine preparations have contained tissue cross-reactive antigens that stimulated the production of antibodies that react with several cardiac and renal antigens (Kaplan and Meyeserian, 1962; Zabriskie and Friemer, 1966; van de Rijn *et al.*, 1977; Dale and Beachey, 1982, 1985; Krisher and Cunningham, 1985; Kraus and Beachey, 1988; Kraus *et al.*, 1989a).

A number of studies have shown that the major virulence factor of group A *Streptococcus pyogenes* is the M protein, which makes up the α-helical fibrils that extend outwards from the surface of these organisms.

This protein has anti-opsonic properties which enable the organism to escape the attack of host phagocytic cells. Antibodies to M protein are protective, by opsonizing the bacteria and leading to phagocytosis and killing by the host (Lancefield, 1962; Whitnack and Beachey, 1982, 1985). Immunity against the type-specific M protein molecule confers protection against subsequent infections by these organisms (Lancefield, 1962). For these reasons, the M protein became an obvious target for vaccine design. However, the discovery of autoimmune epitopes within the covalent structure of the M protein molecule itself (Dale and Beachey, 1985; Kraus and Beachey, 1988; Kraus *et al.*, 1989a) hampered the development of such vaccines. The fear has been that vaccines containing heart cross-reactive determinants may actually trigger rather than prevent rheumatic heart disease. Even though no evidence has been provided that cross-reactive antibodies evoked by M protein vaccines are harmful, the potential for harm has provided the impetus to devise vaccine preparations either free of such determinants or in a form that such epitopes would not be recognized by the immune system of the host.

In this paper, we shall review some of the recent attempts to overcome some of these problems. First we will review the immunology, and molecular biology of streptococcal M protein. We shall then review several different approaches that have been undertaken for the development of safe and effective streptococcal vaccines. Finally, we shall discuss our recent data suggesting the potential impact of human T cell responses to M protein on vaccine design.

Structure of M Protein

Protective immunity against group A streptococci is directed exclusively against the type-specific M protein on the surface of virulent organisms (Lancefield, 1962). Several laboratories have undertaken studies to gain a detailed understanding of the structural and functional aspects of different serotypes of M protein (Beachey *et al.*, 1980; Manjula *et al.*, 1983, 1984; Phillips *et al.*, 1981; Beachey *et al.*, 1983; Hollingshead *et al.*, 1986; Mouw *et al.*, 1988; Miller *et al.*, 1988). The molecular structure of the M protein is reviewed in these proceedings by Dr. Vincent Fischetti, and, therefore, only those points that are relevant to our discussion will be summarized.

Initial studies of the primary structures of various M proteins employed purified pepsin (Beachey *et al.*, 1980) or detergent (Fischetti, 1978) extracts of whole group A streptococci. These purified preparations were subjected to amino acid sequence analysis. Subsequently, the structural genes of several M proteins were cloned, expressed in *Escherichia coli*, and sequenced (Hollingshead *et al.*, 1986; Mouw *et al.*, 1988; Miller *et al.*, 1988). The amino acid sequences of the entire M protein molecules were then deduced from the DNA sequences (Hollingshead *et al.*, 1986; Mouw *et al.*, 1988; Miller *et al.*, 1988). These molecular studies have indicated that the various serotypes of M protein are encoded by allelic genes which exhibit several remarkable features. First, the amino acid sequences of the amino terminal halves of these molecules were highly variable (Hollingshead *et al.*, 1986; Mouw *et al.*, 1988; Miller *et al.*, 1988; Fischetti, 1978; Scott *et al.*, 1986). Second, each of the M protein molecules contain varying lengths of internal sequence repeats, the most striking occurring in type 24 M protein with up to five identical 35 amino acid residue repeats (Beachey *et al.*, 1980; Mouw *et al.*, 1988; Miller *et al.*, 1988). Third, each contains a seven-residue motif with respect to placement of hydrophobic and hydrophilic residues, and this motif is repeated throughout most of the molecule (Hollingshead *et al.*, 1986; Manjula and Fischetti, 1980). This seven-residue periodicity confers an alpha-helical coiled-coil structure upon the M protein molecule (Phillips *et al.*, 1981) (Figure 1). It is the first such conformation to be demonstrated for any surface appendage of bacteria.

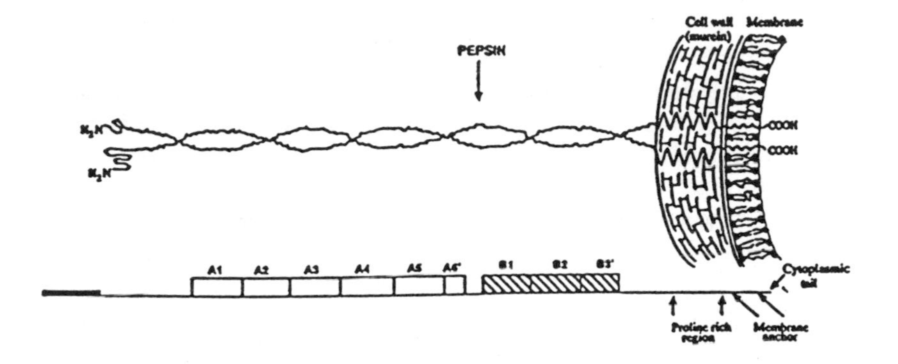

Figure 1. Alpha-helical coiled-coil and repeating covalent structure of streptococcal type 24 M protein. At the bottom is shown the various regions of the covalent structure. The heavy line represents the signal peptide sequence, whereas the thinner line represents the covalent sequence of the mature M protein molecule. The long 35-residue repeating sequences are shown by the blocks. The sequences represented by each of the open boxes are identical to each other but are totally different to the identical repeats represented by the stripped boxes. A proline-rich region followed by a membrane anchor region and a cytoplasmic tail are indicated. Each of these regions is diagrammed above.

Native Polypeptide and Chemically Synthesized Fragment Vaccines

The notion that peptide fragments of M protein might evoke protective immune responses is based on earlier observations that native polypeptide fragments of M protein obtained by hot acid extraction or mild pepsin treatment of the organisms react with type-specific M protein antisera.

These size-heterogenous polypeptides evoked opsonic and protective antibodies in laboratory animals (Fox and Wittner, 1965; Wiley and Wilson, 1960; Wiley and Bruno, 1968; Bergner-Rabinowitz *et al.*, 1972; Dale *et al.*, 1980; Fischetti, 1977). Moreover, immunization of rabbits and humans with a pepsin-extracted 20-kDa fragment of type 5 M protein or a 33-kDa fragment of type 24 M protein evoked opsonic antibodies against the respective serotypes of group A streptococci (Beachey *et al.*, 1977).

Although the native polypeptide fragments described above retained type-specific immunogenicity, such fragments prepared from types 5, 6, and 19 M proteins evoked, in addition, antibodies that cross-reacted with heart tissue. The possibility that one may be able to select peptide regions of the M protein molecules that contain protective, but not tissue cross-reactive epitopes, was then investigated (Beachey *et al.*, 1979, 1981, 1984; Dale *et al.*, 1983; Dale and Beachey, 1986; Beachey and Seyer, 1986). Initial studies were performed with types 5 and 6 M protein, both of which contain heart cross-reactive epitopes (Dale *et al.*, 1983; Dale and Beachey, 1986; Beachey and Seyer, 1986). These were compared with similar studies of type 24 M protein, which lacks such autoimmune epitopes (Beachey *et al.*, 1983, 1981, 1984). We reasoned that the amino terminal ends of the molecules should be readily accessible at the streptococcal surface, and therefore should be the most likely region to contain opsonogenic or protective epitopes. It was also noted that the amino terminal 20-30 residues were predicted to comprise a random coil conformation (Phillips *et al.*, 1981) and thus are less likely to possess epitopes shared with the alpha-helical coiled coil protein, such as cardiac myosin or alpha tropomyosin. Based on these assumptions, 20-residue peptides of types 5 and 6 M proteins were synthesized, conjugated to tetanus toxoid with glutaraldehyde, and used to immunize rabbits (Dale *et al.*, 1983; Beachey and Seyer, 1986). The animals responded uniformly to each of the peptides by producing high titers of antibodies against the respective native M proteins as measured by ELISA. The antisera opsonized types 5 and 6 streptococci in a type-specific fashion but failed to opsonize type 24 organisms. None of the animals developed heart or myosin cross-reactive antibodies (Dale *et al.*, 1983; Beachey and Seyer, 1986). Thus, it appeared that one could select protective regions of M protein that lacked autoimmune epitopes.

To overcome the problem of type-specificity of these amino terminal peptides, attempts were made to prepare more broadly protective vaccines by synthesizing selected peptide regions of several different serotypes of M protein linked in tandem (Beachey *et al.*, 1986, 1987). First, a bivalent vaccine composed of the first 20 residues of type 5 M protein, SM5[1-20], and the 13 carboxy-terminal residues of one of the 35 residue repeats of type 24 M protein, S-CB7[23-35], was synthesized in tandem yielding SM5[1-20]-S-CB7[23-35]C; the carboxyterminal cysteine residue was added for the purpose of coupling to a carrier. The bivalent peptide was found to be highly immunogenic when emulsified in Freund's complete adjuvant without coupling to a carrier. It stimulated high titers of antibodies against the respective native proteins, M5 and M24. The immune sera opsonized both types 5 and 24, but not type 6 streptococci. None of the immunized animals developed autoantibodies against heart tissue or myosin (Beachey *et al.*, 1986).

Next, a trivalent vaccine containing the amino terminal 10 residues of M5, 11 residues of M6, and 12 residues of M24 was synthesized in tandem (Beachey *et al.*, 1987). An important question in these studies was the accessibility to the immune system of the M6 sequence in the middle of the chimeric tripeptide (Figure 2).

This has implications for the future potential of synthetic hybrid vaccines. The tripeptide SM5[1-10]-SM6[1-11]-SM24[1-12]C evoked strong immune responses in four of seven rabbits against each of the three serotypes of M protein. All seven responded briskly to the M6 epitope, the sequence in the middle. None of the animals produced heart cross-reactive

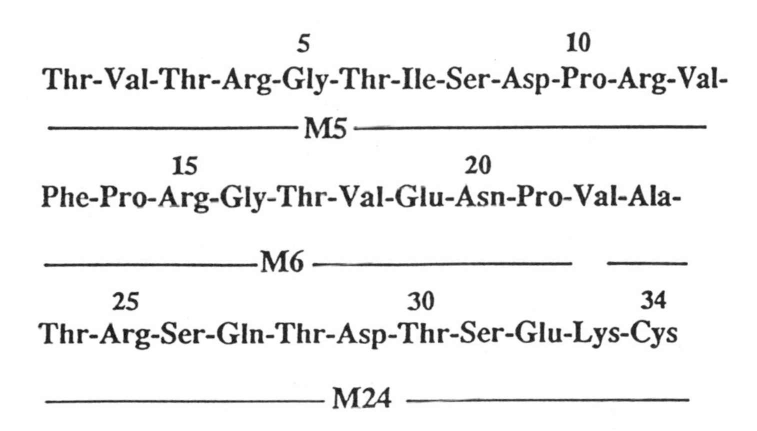

Figure 2. Amino acid sequence of trivalent M protein hybrid peptide.

antibodies (Beachey *et al.*, 1987). These results suggested that the amino terminal, nonhelical regions of various serotypes of M protein possess type-specific protective epitopes and are free of tissue cross-reactivity. That this was not always the case, however, was demonstrated in subsequent studies of the immunogenicity of amino terminal peptides of types 1, 12, and 19 M proteins (Kraus and Beachey, 1988; Kraus *et al.*, 1989a; Newton *et al.*, 1991, 1989). Residues 1-26 of type 1 M protein and 1-25 of type 12 M protein evoked autoimmune responses to renal glomeruli (Kraus and Beachey, 1988; Kraus *et al.*, 1989 a,b), and residues 1-25 of type 19 M protein evoked autoimmune responses to heart tissues (Bronze *et al.*, 1988a). The cross-reactive antibodies raised by the type 1 and 12 amino terminal peptides were shown to react with vimentin, an intermediate filament protein of mesangial cells (Kraus *et al.*, 1989 a,b). The epitope shared between type 1 M protein and vimentin is specified by a tetrapeptide, Ile-Arg-Leu-Arg (Kraus and Beachey, 1988), whereas that shared between type 12 M protein and vimentin is apparently specified by a tertiary structural epitope (Kraus *et al.*, 1989b).

Thus, the selection of non-helical, amino terminal M protein peptides for incorporation into multivalent vaccines is not the complete answer to the problem of auto-immunogenicity of various M protein vaccines. Inasmuch as all of the studies showing autoimmune responses to M protein utilized parenteral vaccine preparations, it was of interest to determine whether such cross-reactions would be elicited by a different route of immunization, namely intranasally or orally. To this end, the protective immune responses of experimental animals to recombinant M protein expressed in attenuated aroA *Salmonella typhimurium* and administered orally was investigated.

Oral Recombinant M Protein Vaccines

In an attempt to overcome the toxicity and autoimmune reactions encountered by the vaccine approaches described above, we undertook studies of the protective and autoimmune responses to M protein cloned and expressed in an attenuated aroA strain of *S. typhimurium* SL3261, which is able to invade mucosal surfaces of the intestinal tract but is unable to cause disease (Poirier *et al.*, 1988). The plasmid pMK207 (Kehoe *et al.*, 1985) containing the M5 structural gene (spm5) was first transformed into an r-m + strain LB5000 of *S. typhimurium*. The plasmid isolated and purified from LB5000 was then used to transform aroA *S. typhimurium* SL3261. The transformed strains LB5000 and SL3261 expressed the entire M protein molecule as shown by Western blot analysis of whole cell lysates (Poirier *et al.*, 1988). The M5 protein was expressed in a stable manner by SL3261 in the absence of antibiotic pressure during repeated subculturing over five days. Furthermore, colonies of SL3261-pMK207 isolated from the liver of a mouse three weeks after oral inoculation continued to express M5 protein (Poirier *et al.*, 1988).

The oral immunization of mice with 10^9 CFU of the attenuated *S. typhimurium* SL3261 expressing M5 protein resulted in both systemic and mucosal immune responses (Poirier *et al.*, 1988). The pooled sera from the immunized mice opsonized type 5 but not type 24 streptococci. None of the serum or salivary antibodies reacted with frozen sections of human heart in immunofluorescent antibody tests and neither reacted with human heart sarcolemmal membranes in Western immunoblots (Poirier *et al.*, 1988).

Based on these results, groups of mice immunized as descibed above were challenged on day 22 with type 5 or 24 streptococci, or virulent *S. typhimurium* SC1344 either intraperitoneally or intranasally. Only the SL3261-pMK207-immunized mice survived either the intraperitoneal or intranasal challenge infections with type 5 streptococci or virulent *S. typhimurium* (Tables 1 and 2). Protection was M-type-specific, as virtually all of the immunized mice challenged with type 24 streptococci died.

Because of the possibility that the immune system of the Balb/c mice used in these studies might not be able to recognize the autoimmune epitopes of type 5M protein, four mice were immunized parenterally with 50 ug doses of pep M5 emulsified in CFA, followed four weeks later with the same dose in saline injected subcutaneously. None of the mice developed heart cross-reactive antibodies, suggesting that these animals may not recognize the autoimmune epitopes.

Because it has been clearly established that rabbits do respond to the M5 autoimmune epitopes, we have initiated studies of the immune responses of New Zealand white rabbits to orally administered recombinant M5 expressed by aroA *S. typhimurium* SL3261. In preliminary studies, three rabbits fed 10^9 organisms on days one and six developed opsonic type 5 antibodies four weeks after the initial immunizing dose even though the ELISA titers of serum antibodies were relatively low, ranging from 1:400 to 1:3200 (Poirier, unpublished data). None of the immune sera reacted with heart tissue either by immunofluorescence tests of frozen sections or by Western blot analysis of SDS-solubilized myocardial tissue (unpublished). These early results suggest that the stimulation of autoimmune antibodies is dependent either on the route of immunization or on the dose of M protein administered. We had previously shown that only one of three rabbits receiving 100 ug doses of M5 protein developed heart cross-reactive antibodies, whereas virtually all rabbits receiving 300 ug doses of antigen developed such antibodies. We have estimated that the dose of M protein per 10^9 *S. typhimurium* is approximately 5 ug, which is far less than usually required for optimal responses by the parenteral route.

It is clear from the oral vaccine studies in mice that protection was virtually complete against both parenteral and mucosal challenge infections even though the serum antibody titers against M5 protein were relatively low; the pooled serum titer was only 400 at three weeks when the mice were challenged. Similarly low titers of M protein-specific antibodies were obtained in mice immunized intranasally with purified type 24 M protein (pep M24) or with whole heat-killed type 24 streptococci (Bronze *et al.*, 1988b). Yet these animals were protected against challenge infections with group A streptococci (Bronze *et al.*, 1988b). The reason for the high degree of protection in face of a low humoral antibody response is not clear, although it is clear that the animals respond briskly to booster doses of purified M protein vaccine administered subcutaneously. The IgG antibody titer was boosted from 400 to 12,800 in pooled sera obtained from mice 60 hours after a booster injection of 50 ug of M5 protein. It is possible that the challenge dose of live bacteria induces a similar booster response that may account for the high degree of protection.

Table 1

Challenge by intra-peritoneal injection of M type 5 and type 24 streptococci of *S. typhimurium* SL1344 in BALB/c mice immunized orally with live aroA *S. typhimurium* transformed with pMK207 expressing type 5 M protein

		Survival[b]	
Challenge organisms	Dose[a]	Unimmunized	Immunized
M5 Streptococci	1.7×10^4	2/4	4/5
(Smith strain)	1.7×10^5	0/4	5/5
	1.7×10^6	0/4	5/5
M24 Streptococci	1.6×10^3	1/3	1/3
(Vaughn strain)	1.6×10^4	0/3	0/3
	1.6×10^5	0/3	0/3
S. typhimurium	7.3×10^1	0/3	3/3
(Strain SL1344)	7.3×10^2	0/3	3/3
	7.3×10^3	0/3	3/3

[a] Dose was determined as the CFU administered.

[b] Survival was recorded as the number of surviving mice divided by the number of mice challenged.

Data from Poirier *et al*. (1988).

Table 2

Challenge by intra-nasal inoculation of M types 5 and 24 streptococci or *S. typhimurium* SL1344 in BALB/c mice immunized orally with live aroA *S. typhimurium* transformed with pMK207 expressing type 5 M protein

		Survival[b]	
Challenge organisms	Dose[a]	Unimmunized	Immunized
M5 Streptococci (Smith strain)	3.9 X 10^7	0/4	6/6
M24 Streptococci (Vaughn strain)	3.5 X 10^7	0/4	0/6
S. typhimurium (Strain SL1344)	4.3 X 10^4	0/4	6/6

[a] Challenge dose was determined as the CFU administered intra-nasally and given 13 wk after the initial immunization (i.e., day 1).

[b] Survival was recorded as the number of surviving mice over the number of mice challenged.

Data from Poirier *et al.* (1988).

Expression and Immunogenicity of M Protein Epitope Inserted in *Salmonella* flagellin

It has now been demonstrated that attenuated strains of *Salmonella spp.* may have a role as an effective, general vaccine delivery systems because they can be used to express a variety of foreign antigens intact, or as peptide epitopes, and carry them to the host's immune system. In an attempt to optimize this capability, we explored the possibility that the location of the M protein epitope(s) may enhance the efficacy of the vaccine preparation. by maximizing immunogen delivery (Newton *et al.*, 1991).

Work from Dr. Stocker's laboratory (reviewed in this proceedings) demonstrated that flagellar filaments can be used as carriers of unrelated epitopes by insertion of epitope-specifying oligonucleotides at a restriction site in a region of the gene coding for flagellar-antigen determinants (Newton *et al.*, 1989). Therefore, in collaboration with Drs. Newton and Stocker, we synthesized a 48 base-pair oligonucleotide specifying the N-terminal 15 amino acids of M protein of *Streptococcus pyogenes* type 5 (plus a CTA codon, to terminate translation of genes with the insert in reverse orientation) and inserted it by blunt-end ligation at the site of the 48 base-pair EcoRV deletion in the *Salmonella* flagellin gene in plasmid pLS408 (Newton *et al.*, 1989). The resulting plasmid was transferred from *E. coli* via a restriction-negative *S. typhimurium* into an aromatic-dependent flagellin-negative live-vaccine strain of *S. dublin*, to produce strain SL7127, which was motile. Expression of the inserted epitope in flagellin and its exposure at the flagellar filament surface were shown by immunobloting and by the reaction of flagellate bacteria (immobilization, immunogold-labelling) with antibody raised by injection of the corresponding synthetic peptide, S-M5[1-15] (Figure 3).

Rabbits immunized by injection of the live-vaccine strain with flagella composed of the chimeric flagellin or by injection of concentrated flagella from such bacteria developed antibodies reactive in ELISA test with peptide S-M5[1-15] and with the large peptic-digest peptide (pepM5). These antibodies were opsonophagocytic for type 5 streptococci. BALB/c mice given five doses i.p., each 5 X 10^6 live SL7127, over an 8 week period developed titers of ca.1,000 for the M5-specific peptides and opsonophagocytic activity for type 5 but not for type 24 streptococci (Table 3). Four of the 5 mice given strain SL7127, with insert, survived the M5 challenge but none of the five challenged with the type 24 strain survived (Table 4). In contrast, all of the five mice given the control strain, without insert, died after challenge with type 5 streptococci or type 24 streptococci. Thus mice can be protected against a *Streptococcus pyogenes* type 5 challenge by immunization with a *Salmonella* live-vaccine with flagella made of flagellin with an insert carrying a protective epitope of M5 protein but without the cross-reactive epitopes of the complete protein (Newton *et al.*, 1991).

It remains to be seen whether these live vaccine strains with flagella containing larger inserts of the M protein molecule that contain both type - specific and non-specific epitopes can be used to protect animals without evoking autoreactive antibodies or self-reactive T lymphocytes. The previous studies using attenauted *S. typhimurium* as a delivery vehicle demonstrated that although the entire M protein molecule is expressed, the autoimmune epitopes are not recognized in test animals. This could be attributed to the lower dose of M protein required to produce a protective immune response when recombinant vaccines are delivered by an attenuated bacterial vehicle. Whether or not these live vaccines will behave in a similar manner in humans remains to be tested. However, before human trials can be conducted, the response of human T cells to M protein has to be evaluated.

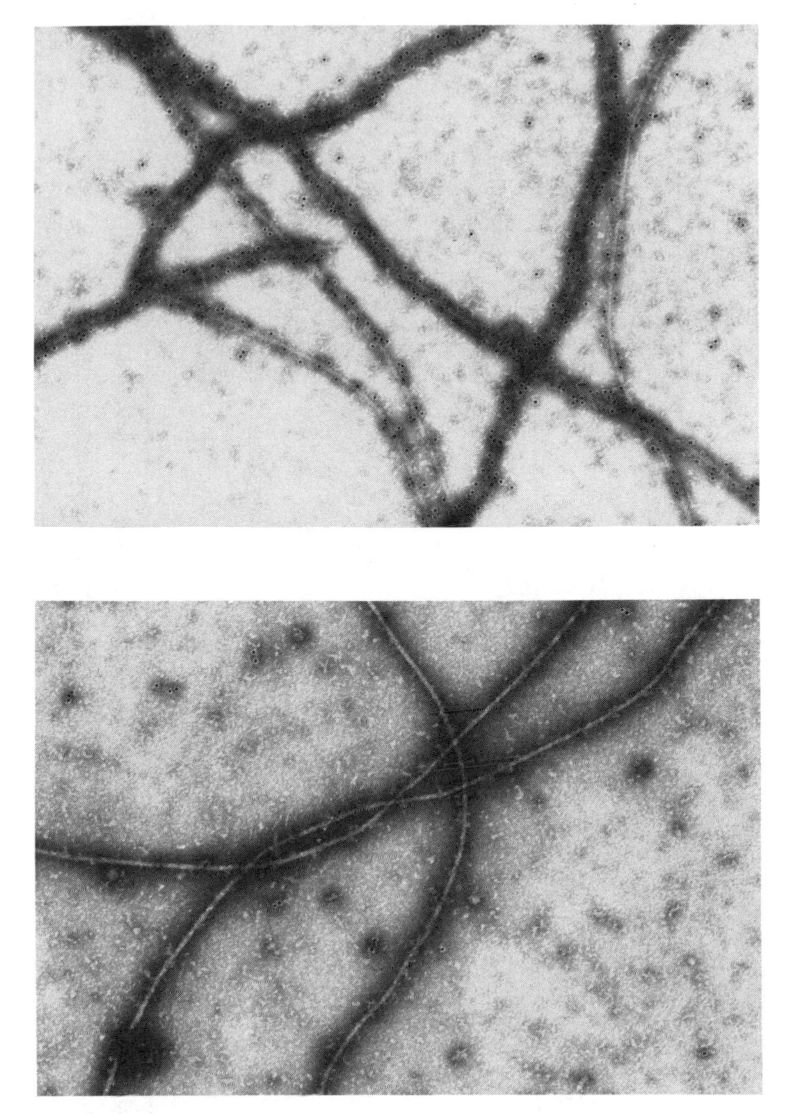

Figure 3. Immunogold labeling of the M5 epitope at the surfae of the flagella. (A) Strain SL7127 (with the M5 insert) and (B) Strain SL5930 (without the insert) were labeled by treatment with anti-S-M5(1-15) and gold-conjugated goat anti-rabbit antibody. Data from Newton *et al.* (1991).

Table 3

Immune Responses in BALB/c Mice Immunized with Live *S. dublin* SL7127 and SL5930

	SL7217-inmmunized mice				SL5930-immunized mice			
	ELISA titer[a]			% opsonization	ELISA titer			% opsonization
Test Serum	S-M5(1-15)C	PepM5	SL5930	T5(Smith)	S-M5(1-15)C	PepM5	SL5930	T5(Smith)
Pre-immune	<100	<100	<100	0	<100	<100	<100	0
2wk	200	400	3,200	6	<100	<100	3,200	0
4wk	800	3,200	12,800	52	<100	<100	6,400	0
6wk	6,400	6,400	25,600	96	<100	100	12,800	0
8wk	6,400	12,800	12,800	92	100	100	12,800	0

Groups of 5 mice were immunized with either strain SL7127 or SL5930. Each dose, given i.p., comprised 1-2 X 10^6 live bacteria. Pooled sera from each group of 5 mice, bled before the first and each subsequent (booster) inoculation, were titrated by ELISA with synthetic peptide S-M5(1-15)c, pep M5 or bacteria of strain SL5930 (no insert in flagellin) as coating antigen. Each serum pool was also tested for opsonization of *S. pyogenes* type 5 (strain Smith).
Data from Newton *et al.* (1991).

Table 4

Challenge by intra-peritoneal injection of M type 5 and type 24 streptococci or *S. dublin* SL5608 in BALB/c mice immunized with live *S. dublin* SL7217 or SL5930

	Survival of Mice immunized with Strain		
Challenge [a] organisms	SL7217	SL593	unimmunized
M5 Streptococci (Smith strain)	4/5	0/5	ND
M24 Streptococci (Vaughn strain)	0/5	0/5	ND
S. dublin (Strain SL5608)	5/5	5/5	0/5

[a] Fifteen days after the last vaccination, mice were challenged with either 1.6×10^6 type M5 streptococci, 2.2×10^5, or 6.3×10^3 *S. dublin*.

N D : Not done

Data from Newton *et al.* (1991).

Cell Mediated Immune Responses to M Protein and its Impact on Vaccine Design

Although certain streptococcal products are extremely toxic and others give rise to immune responses against host tissues, especially the heart, none of the toxic products have been shown to be directly related to pathogenesis of rheumatic fever or cardiac damage, and the pathogenic significance of heart cross-reactive immune responses has remained unclear. In fact, many patients who contract streptococcal pharyngitis develop antibodies against cardiac tissue and yet most have no evidence of rheumatic fever or cardiac injury (Lancefield, 1962).

Several lines of evidence indicate a possible role for cell-mediated immunity in the pathogenesis of post-streptococcal diseases. The presence of lymphocytic infiltrates in the hearts of acute rheumatic fever patients (Raizada *et al.*, 1983) and of circulating cytotoxic T lymphocytes (CTL) in their blood, directed towards cardiac cells (Hutto and Ayoub, 1980) strongly suggested a role for T cells in rheumatic carditis. That M protein is involved in these responses has been suggested by several studies showing that human T cells respond briskly to purified, pepsin-extracted fragments of M protein (pep M) (Beachey *et al.*, 1979, 1981, 1969; Dale *et al.*, 1981; Dale and Beachey, 1987). The response was not merely due to prior exposure to streptococcal infections, inasmuch as cord blood lymphocytes responded in a similar fashion (Dale *et al.*, 1981). This suggested that M protein may be a polyclonal T cell mitogen; however M protein does not behave like a typical mitogen. The response of resting T cells to M protein was consistently lower than the response to known T cell mitogens, such as PHA and Con A. In addition, M protein triggered biochemical reactions in T cells that were different from those triggered by PHA (Kotb *et al.*, 1987; Kotb and Beachey, 1989), and furthermore different M protein serotypes stimulated phenotypically and functionally distinct T cell subpopulations (Table 5). Certain M protein serotypes (e.g. types 5 and 6) induce the differentiation of CD8 and CD4 cytotoxic T lymphocytes (CTL) (Dale and Beachey, 1987), whereas other serotypes (e.g. type 24) induce CD8 suppressor T cell activity (Kotb *et al.*, 1989a).

It became clear to us that the mechanism by which M protein stimulates T cells needed to be understood and the mitogenic T cell epitopes of the M protein molecule needed to be discerned. Therefore, we conducted thorough studies of the cellular and molecular requirement for T cell activation by M protein. Pepsin extracted type 5 M protein (pep M5) induced T cell proliferation is accessory cell-dependent; however, the presentation of pep M5 to T cells does not appear to be MHC-restricted, since allogeneic cells were as effective as autologous cells in supporting human T cell proliferation in response to pep M5 (Kotb *et al.*, 1989b; Beachey *et al.*, 1990; Tomai *et al.*, 1990). Unlike PHA, M protein-induced T cell stimulation required the expression of class II molecules by the antigen presenting cells (APC). This was shown in experiments in which mouse L cells, transfected and expressing high densities of both HLA class II a and b chains were used as APC (Tomai *et al.*, 1990). Fibroblasts transfected with HLA-DR, DP, and DQ were all capable of presenting pep M5 to purified human T cells. In contrast, fibroblasts transfected with the neomycin resistance gene alone (L66) did not present pep M5, but did provide the necessary component(s) for stimulation by PHA (Figure 4).

Monoclonal antibodies to either HLA-DR or HLA-DR, DQ inhibited the presentation of M protein but did not significantly affect the stimulation of T cells by PHA (Tomai *et al.*, 1990). These results indicate that M protein can be presented by HLA class II elements regardless of their isotype or haplotype. Moreover, binding studies demonstrated that pep M5 associates directly to MHC class II molecules on antigen presenting cells, and that unlike conventional antigens, the processing of M protein by the APC is not required. The unprocessed protein stimulates both CD4 or CD8 T cell subsets via interacting with the T cell receptor (TCR) (Table 6). Of particular importance is the finding that pep M5 does not stimulate all T cells because it preferentially stimulates T cell subsets bearing $V\beta 2$, $V\beta 4$ and $V\beta 8$ sequences in the a β portion of their TCR (Tomai *et al.*, 1990). We are investigating the region within the M protein molecule responsible for this activity.

Table 5

Suppression of Pep M6-induced CTL activity by Pep M24 stimulated PB lymphocytes

		%specific ^{51}Cr release[b]	
Pre-incubation[a]	stimulation	GHC	K562
None	Medium	4	35
None	Pep M6	32	35
Medium	Pep M6	24	26
Pep M24	Pep M6	6	41

[a] PB lymphocytes were cultured in medium alone (control) or with 10 ug/ml pep M24. After 3 days in culture, the cells were washed and 0.25 x 10^5 control or pep M24 stimulated cells were mixed with 10^5 fresh autolougous cells and then stimulated with 10 ug/ml pep M6.

[b] After 5 days in culture the cells werc tested for cytotoxicity against GHC and K562 cells at an effector : target ratio of 100.

Data from Kotb *et al.* (1989a).

Overlaping synthetic peptides representing various regions of the M protein molecule and consisting of up to 35 amino acids failed to induce T cell stimulation, whereas larger peptides of about 85 amino acids were able to stimulate T cells but only in a few individuals (Kotb, unpublished). These data indicate that M protein does not stimulate T cells via classical T cell epitopes; rather, M protein is a superantigen (Kotb *et al.*, 1989 b,c; Beachey *et al.*, 1990; Tomai *et al.*, 1990; Majumdar *et al.*, 1991) which requires larger domains to bind to both HLA class II molecules on the APC and the $V\beta$ chain of the TCR on T cells.

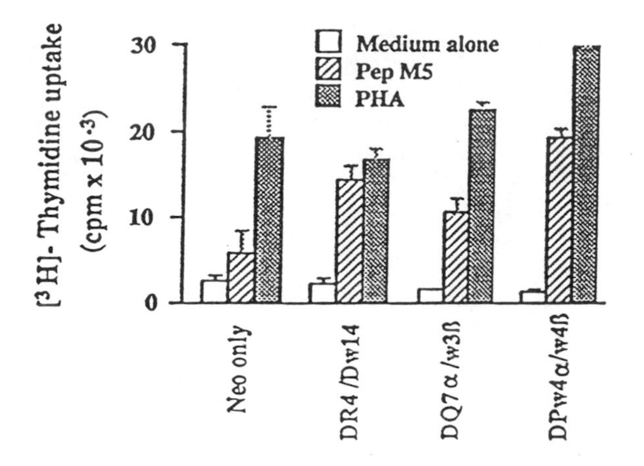

Figure 4. Presentation of pep M5 by HLA class II transfected mouse fibroblasts. T cells (10^5) were cultured with or without 2 X 10^4 mitomycin C-treated mouse fibroblasts transfected with: neomycin resistance gene only (L66), HLA-DR (L165.1), HLA-DQ (L54.5), or HLA-DP(L25.4). These fibroblast were a generous gift of Dr. Robert Karr. Cultures were incubated for 3 days in medium alone or in the presence of pep M5 (5ug/ml), or PHA (5ug/ml). Proliferation was assessed by [3H]-thymidine incorporation. Data from Tomai *et al.* (1990).

In light of these new findings, streptococcal vaccines containing M protein epitopes should be tested for their effect on cell mediated immune responses prior to their administration into humans. Interestingly, the superantigenicity of M protein is confined to human T cells. Lymphocytes from other animal species recognize the M protein as a regular antigen, and not as a superantigen. This constitutes a problem because we will not be able to test the effects of M protein vaccine preparations on T cells in animals and extrapolate the results to humans. Studies using human cells will have to be employed, but these can only be performed *in vitro*. To overcome these obstacles, we will attempt to develop a human transgenic animal model which will enable us to test such vaccine preparations. It is clear therefore, that further studies of the T cell responses to M protein are required, not only to understand its role in the pathogenesis of rheumatic fever and rheumatic heart disease, but also to devise safe and effective vaccines for the prevention of these diseases.

Table 6

Analysis of T cell subpopulations in Pep M5 stimulated PB lymphocytes

Culture	CD8	CD4/2H4	CD4/4B4
Control	15	42	15
	17	39	15
Pep M5	26	43	29
	30	40	27

PB lymphocytes were cultured for 4 days with medium alone (control) or with 10 ug/ml pep M5. T cell subpopulations were analyzed by flow cytometry using flourescent-conjugated monoclonal antibodies directed towards the indicated T cell surface markers. The data are from two separate cultures, presented as % positive cells.

Summary

For more than 50 years, attempts have been made to develop safe and effective vaccines against group A streptococcal infections. The major problems have been a high degree of toxicity of almost any vaccine preparation tested in humans. Perhaps the most bothersome problem has been the finding of autoimmune epitopes within the covalent structure of the M protein molecule itself. Thus efforts were directed towards the development of vaccine preparations that are either free of such epitopes or that are introduced in a form that makes these epitopes cryptic to immune recognition. Two approaches reviewed above for the development of such vaccines hold promise. The first is the development of multivalent synthetic peptide vaccines composed of selected protective regions of M protein that lack autoimmune epitopes. Additional work is required, however, to determine the efficacy of such vaccines in humans. The second approach which holds promise is the development of an oral vaccine composed of attenuated *Salmonella spp*. expressing the cloned M protein of group A streptococci. Thus far it appears that although the entire M protein molecule is expressed, the autoimmune epitopes are not recognized either in mice or in rabbits. The latter may be due to the much lower dose of M protein required to produce a protective immune response when recombinant vaccines are delivered by an attenuated bacterial vehicle. The advantage of the cloned preparations is that various truncated forms or chimeric constructions containing protective epitopes from several different serotypes of M protein can readily be made by genetic manipulations. It is now necessary to extend the animal studies with recombinant M protein vaccines to humans. However, before we embark on these human trials, the cell mediated immune responses to M protein will have to be better understood. This is particularly important in light of clinical evidence indicating that T cells play an important role in the pathogenesis of post streptococcal autoimmune diseases. It is clear therefore, that further studies of the cellular responses to M protein may provide new insights not only into the pathogenesis of rheumatic fever and rheumatic heart disease but also into the formulation of safe and effective vaccines for the primary prevention of these diseases.

Acknowledgements

This review was prepared by the late Dr. Beachey and myself in 1989, in response to a request from the Steering Committee on Encapsulated Bacteria of the WHO Programme for Vaccine Development. It was presented as a guidance document on group A streptococcus and prospects for a vaccine. The WHO committee was kind enough to permit the publication of this updated review in this proceedings. The research efforts of the authors were supported by research funds from the Veterans Administration and by research grants from the National Institutes of Health, AI-10085 and AI-13550 to E. H. Beachey, and GM-3580 to M. Kotb.

References

Beachey, E.H., Alberti, H., and Stollerman, G.H., 1969, Delayed hypersensitivity to purified streptococcal M protein in guinea pigs and in man. *J. Immunol.* 102:42.

Beachey, E.H., Gras-Masse, H., Tartar, A., Jolivet, M., Audibert, F., Chedid, L., and Seyer, J.M., 1986, Opsonic antibodies evoked by hybrid peptide copies of types 5 and 24 streptococcal M proteins synthesized in tandem. *J. Exp. Med.* 163:1451.

Beachey, E.H., Majumdar, G., Tomai, M., and Kotb, M., 1990, Molecular aspects of autoimmune responses to streptococcal M protein. In : Advances in Host Defense Mechanisms (Gallin and Fauci, eds.) Raven Press, p. 83.

Beachey, E.H., and Seyer, J.M., 1986, Protective and nonprotective epitopes of chemically synthesized peptides of the NH2-terminal region of type 6 streptococcal M protein. *J. Immunol.* 136:2287.

Beachey, E.H., Seyer, J.M., and Dale, J.B., 1987, Protective immunogenicity and T lymphocyte specificity of a trivalent hybrid peptide containing NH2-terminal sequences of types 5, 6, and 24 streptococcal M proteins synthesized in tandem. *J. Exp. Med.* 66:647.

Beachey, E.H., Seyer, J.M., Dale, J.B., and Hasty, D.L., 1983, Repeating covalent structure and protective immunogenicity of native and synthetic polypeptide fragments of type 24 streptococcal M protein. *J. Biol. Chem.* 258:13250.

Beachey, E.H., Seyer, J.M., Dale, J.B., Simpson, W.A., and Kang, A.H., 1981, Type-specific protective immunity evoked by synthetic peptide of *Streptococcus pyogenes* M protein. *Nature* (London) 292:457.

Beachey, E.H., Seyer, J.M., and Kang, A.H., 1980, Primary structure of protective antigens of type 24 streptococcal M protein. *J. Biol. Chem.* 255:6284.

Beachey, E.H., Stollerman, G.H., Chiang, E.Y., Chiang, T.M., Seyer, J.M., and Kang, A.H., 1977, Purification and properties of M protein extracted from group A streptococci with pepsin. Covalent structure of the amino terminal region of type 24 M antigen. *J. Exp. Med.* 145:1469.

Beachey, E.H., Stollerman, G.H., Johnson, R.H., Ofek, I., and Bisno, A.L., 1979, Human immune response to immunization with a structurally defined polypeptide fragment of streptococcal M protein. *J. Exp. Med.* 150:862.

Beachey, E.H., Tartar, A., Seyer, J.M., and Chedid, L., 1984, Epitope-specific protective immunogenicity of chemically synthesized 13-, 18-, and 23 residue peptide fragments of streptococcal M protein. *Proc. Natl. Acad. Sci. USA* 81:2203.

Bergner-Rabinowitz, S., Ofek, I., and Moody, M.D., 1972, Cross-protection among serotypes of group A streptococci. *J. Infect. Dis.* 125:339.

Bronze, M., Beachey, E.H., and Dale, J.B., 1988a, Opsonic and heart-cross-reactive epitopes located within the amino-terminus of type 19 streptococcal M protein. *J. Exp. Med.* 167:1849.

Bronze, M., McKinsey, D., Beachey, E.H., and Dale, J.B., 1988b, Protective immunity evoked by locally administered group A streptococcal vaccines in mice. *J. Immunol.* 141:2767.

Dale, J.B., and Beachey, E.H., 1982, Protective antigenic determinant of streptococcal M protein shared with sarcolemmal membrane protein of human heart. *J. Exp. Med.* 156:1165.

Dale, J.B., and Beachey, E.H., 1985, Multiple heart cross-reactive epitopes of streptococcal M proteins. J. Exp. Med. 161:113.

Dale, J.B., and Beachey, E.H., 1986, Localization of protective epitopes of the amino terminus of type 5 streptococcal M protein. *J. Exp. Med.* 163:1191.

Dale, J.B., and Beachey, E.H., 1987, Human cytotoxic T lymphocytes evoked by group A streptococcal M proteins. *J. Exp. Med.* 166:1825.

Dale, J.B., Ofek, I., and Beachey, E.H., 1980, Heterogeneity of type-specific and cross-reactive antigenic determinants within a single M protein of group A streptococci. *J. Exp. Med.* 151:1026.

Dale, J.B., Seyer, J.M., and Beachey, E.H., 1983, Type-specific immunogenicity of a chemically synthesized peptide fragment of type 5 streptococcal M protein. *J. Exp. Med.* 158:1727.

Dale, J.B., Simpson, W.A., Ofek, I., and Beachey, E.H., 1981, Blastogenic responses of human lymphocytes to structurally defined polypeptide fragments of streptococcal M protein. *J. Immunol.* 126:1499.

Fischetti, V.A., 1977, Streptococcal M protein extracted by nonionic detergent. II. Analysis of the antibody response to the multiple antigenic determinants of the M protein molecule. *J. Exp. Med.* 146:1108.

Fischetti, V.A., 1978, Streptococcal M protein extracted by nonionic detergent. III. Correlation between immunological cross-reactions and structural similarities with implications for antiphagocytosis. *J. Exp. Med.* 147:1771.

Fox, E.N., and Wittner, M.K., 1965, The multiple molecular structure of the M proteins of group A streptococci. *Proc. Natl. Acad. Sci. USA* 54:1118.

Hollingshead, S.K., Fischetti, V.A., and Scott, J.R., 1986, Complete nucleotide sequence of type 6 M protein of the group A streptococcus: repetitive structure and membrane anchor. *J. Biol. Chem.* 261:1677.

Hutto, J.H., and Ayoub, E.M., 1980, Streptococcal Diseases and the Immune Response, pp.733. Academic Press, New York.

Kaplan, M.H., and Meyeserian, H., 1962, An immunologic cross-reaction between group A streptococcal cells and human heart. *Lancet I*:706.

Kehoe, M., Poirier, T.P., Beachey, E.H., and Timmis, K.N., 1985, Cloning and genetic analysis of serotype 5 M protein determinant of group A streptococci: evidence for multiple copies of the M5 determinant in *Streptococcus pyogenes* genome. *Infect. Immun.* 48:190.

Kholy A, E.L., Rotta, J., Wannamaker, L.W., Strasser, V., Bytchenko, B., Ferreira, W., Houang, L., and Liisberg, E., 1978, *Bull. W.H.O.* 56:887.

Kotb, M., and Beachey, E.H., 1989, Serine and tyrosine phosphorylation of a 28- and 35-kDa proteins of human lymphocytes stimulated by streptococcal M proteins. *Biochem. Biophys. Res. Comm.* 158: 803.

Kotb, M., Courtney H.S., Dale, J.B., and Beachey, E.H., 1989a, Cellular and biochemical responses of human T lymphocytes stimulated with streptococcal M proteins. *J. Immunol.* 142:966.

Kotb, M., Dale, J.B., and Beachey, E.H., 1987, Stimulation of S-adenosylmethionine synthetase in human lymphocytes by streptococcal M protein. *J. Immunol.* 139:202.

Kotb, M., Majumdar, G., Tomai, M., and Beachey, E.H., 1989c, Accessory cell-independent stimulation of human T cells by superantigens. *J. Immunol.* 145:1332.

Kotb, M., Tomai, M., Majumdar, G., and Beachey, E.H., 1989b, Streptococcal M protein is a Superantigen. *Cell Biology*. 109: 29a.

Kraus, W., and Beachey, E.H., 1988, Renal autoimmune epitope of group A streptococci specified by M protein tetrapeptide Ile-Arg-Leu-Arg. *Proc. Natl. Acad. Sci. USA* 85:4516.

Kraus, W., Ohyama, K., Snyder, D.S., and Beachey, E.H., 1989a, Autoimmune sequence of streptococcal M protein shared with the intermediate filament protein, vimentin. *J. Exp. Med.* 169:481.

Kraus, W., Seyer, J.M., and Beachey, E.H., 1989b, Vimentin cross-reactive epitope of type 12 M streptococcal M protein. *Infect. Immun.* 57:2457.

Krisher, K., and Cunningham, M.W., 1985, Myosin: a link between streptococci and heart. *Science* 227:413.

Lancefield, R.C., 1962, Current knowledge of type-specific M antigens of group A streptococci. *J. Immunol.* 89:307.

Majumdar, G., Beachey, E.H., Tomai, M., and Kotb, M., 1991, Differential signal requirements in T cell activation by mitogen and superantigen. *Cell. Signalling* 2: 521.

Manjula, B.N., Acharya, A.S., Mische, S.M., Fairwell, T., and Fischetti, V.A., 1984, The complete amino acid sequence of a biologically active 197-residue fragment of M protein isolated from type 5 group A streptococci. *J. Biol. Chem.* 259:3686.

Manjula, B.N., and Fischetti, V.A., 1980, Tropomyosin-like seven residue periodicity in three immunologically distinct streptococcal M proteins and its implications for the antiphagocytic property of the molecule. *J. Exp. Med.* 151:695.

Manjula, B.N., Mische, S.M., and Fischetti, V.A., 1983, Primary structure of streptococcal pep M5 protein: absence of extensive sequence repeats. *Proc. Natl. Acad. Sci. USA* 80:5475.

Miller, L., Gray, L., Beachey, E.H., and Kehoe, M., 1988, Antigenic variationamong group A streptococcal M proteins: nucleotide sequence of the serotype 5 M protein gene and its relationship with genes encoding types 6 and 24 M proteins. *J. Biol. Chem.* 263:5668.

Mouw, A.R., Beachey, E.H., and Burdett, V., 1988, Molecular evolution of streptococcal M protein: cloning and nucleotide sequence of the type 24 M-protein gene and relation to other genes of *Streptococcus pyogenes* SL3261. *J. Bacteriol.* 170:676.

Newton, S.M., Kotb, M., Poirier, T.P., Stocker, B.A.D., and Beachey, E.H., 1991, Expression and immunogenicity of streptococcal M protein epitope inserted in *Salmonella* flagellin. *Infect. Immun.* 59: 2158.

Newton, S.M.C., Jacob, C.O., and Stocker, B.A.D., 1989, Immune response to cholera toxin epitope inserted in *Salmonella* flagellin. *Science* 244:70.

Phillips, G.N., Flicker, P.F., Cohen, C., Manjula, B.N., and Fischetti, V.A., 1981, Streptococcal M protein: alpha-helical coiled-coil structure and arrangement on the cell surface. *Proc. Natl. Acad. Sci. USA* 78:4689.

Poirier, T.P., Kehoe, M., and Beachey, E.H., 1988, Protective immunity evoked by oral administration of attenuated aro A *Salmonella typhimurium* expressing cloned streptococcal M protein. *J. Exp. Med.* 168:25.

Raizada, V., *et al.*, 1983, Tissue distribution of lymphocytes in rheumatic heart valves as defined by monoclonal anti-T cell antibodies. *Am. J. Med.* 74:90.

Scott, J.R., Pulliam, W.M., Hollingshead, S.K., and Fischetti, V.A., 1986, Relationship of M protein genes in group A streptococci. *Proc. Natl. Acad Sci USA* 82:1822.

Stollerman, G.H., 1975, Rheumatic Fever and Streptococcal Infection. Grune and Stratton, New York.

Tomai, M., Kotb, M., Majumdar, G., and Beachey, E.H., 1990, Superantigenicity of streptococcal M protein. *J. Exp. Med.* 172:359.

van de Rijn, I., Zabriskie, J.B., and McCarty, M., 1977, Group A streptococcal antigens cross-reactive with myocardium. Purification of heart-reactive antibody and isolation and characterization of the streptococcal antigen. *J. Exp. Med.* 146:579.

Whitnack, E., and Beachey, E.H., 1982, Antiopsonic activity of fibrinogen bound to M protein on the surface of group A streptococci. *J. Clin. Invest.* 69:1042.

Whitnack, E., and Beachey, E.H., 1985, Inhibition of complement-mediated opsonization and phagocytosis of *Streptococcus pyogenes* by D fragments and fibrin bound to cell-surface M protein. *J. Exp. Med.* 162:1983.

Wiley, G.G., and Bruno, P.N., 1968, Cross-reactions among group A streptococci. I. Precipitin and bactericidal cross-reactions among types 33, 41, 43, 52 and Ross. *J. Exp. Med.* 128:959.

Wiley, G.G., and Wilson, A.T., 1960, The occurrence of two M-antigens in certain group A streptococci related to type 14. *J. Exp. Med.* 113:451.

Zabriskie, J.B., and Friemer, E.H., 1966, An immunological relationship between the group A streptococcus and mammalian muscle. *J. Exp. Med.* 124:661.

Protection against Streptococcal Mucosal Colonization

Vincent A. Fischetti, Debra E. Bessen, Olaf Schneewind, and Dennis E. Hruby

Introduction

In the United States alone, 25 to 35 million cases of group A streptococcal infections are reported annually, which primarily afflict school-age children. This high incidence of streptococcal disease, with its potential sequelae of rheumatic fever and acute glomerulonephritis, provides the motivation for efforts to develop an effective and safe vaccine to prevent streptococcal-related diseases. Since the observation by Lancefield that antibodies to the M protein have the capacity to opsonize streptococci in preparation for phagocytic clearance (Lancefield, 1928), the M protein has been a prime candidate for a vaccine to prevent group A streptococcal infections (Bessen and Fischetti, 1990c). However, the ability of the organism to change the antigenic appearance of the M molecule (85 M protein types have now been identified) has hindered the development of this protein for use as a vaccine to protect against all existing group A streptococcal serotypes.

M Protein Structure and Immunochemistry

Analysis of the structure and antigenic determinants of the M molecule reveals that it forms an α−helical coiled-coil structure extending nearly 60 nm from the cell surface (Phillips *et al.*, 1981). Based on structural and serological data, the N-terminal region (distal from the cell wall) is hypervariable and contains the antigenically distinct determinants responsible for type specificity (Jones and Fischetti, 1988). Antigenic homology among M molecules of different serotypes progressively increases at sites which are more proximal to the cell wall with anywhere from 60% to 100% sequence identity at exposed regions close to the cell surface (Jones *et al.*, 1985; Scott *et al.*).

Antibody directed to the type-specific determinants have been found to protect against clinical infection by group A streptococci (Lancefield, 1962). Type-specific serum IgG neutralizes the antiphagocytic property of M protein and initiates opsonophagocytosis. In contrast, antibodies directed to conserved regions of M protein fail to stimulate opsonophagocytosis despite their ability to fix complement (via the classical pathway) as effectively as N-terminal directed antibodies (Jones and Fischetti, 1988). Perhaps, the finding that M protein binds factor H, a regulatory protein of the alternative complement pathway, may explain the inability of the cross-reactive antibodies to initiate phagocytosis (Horstmann *et al.*, 1988).

Type-specific opsonizing antibodies appear to play a central role in providing protection against streptococcal infection after invasion of the mucosal surface. However, the importance of the antiphagocytic property of M protein during the initial stages of infection is less well understood. In limited M protein vaccine trials, humans immunized at an intranasal site displayed lower rates of both nasopharyngeal colonization and clinical illness following streptococcal challenge as compared to the placebo group (Polly *et al.*, 1975). In contrast, individuals immunized subcutaneously exhibited a decrease in clinical illness only, and had no change in the level of nasopharyngeal colonization. These studies suggest that local immune factors play a key role in providing protection against initial infection by group A streptococci.

The Role of M Protein in Streptococcal Adherence

Group A streptococci colonize the nasopharyngeal mucosa of humans, leading to an inflammatory response followed by clinical illness or an asymptomatic carrier state. It was shown (Ellen and Gibbons, 1972) that streptococci bearing the M protein molecule on their surface can adhere better to epithelial cells *in vitro* than M-deficient organisms. There is also strong evidence that lipoteichoic acid (LTA) secreted from the streptococcus mediates direct contact with the epithelial cell surface. M protein or another surface molecule might facilitate this process by anchoring and orienting the LTA adhesin so that the lipid moiety is able to make contact with the epithelial cell (Beachey *et al.*, 1983). Studies also suggest that M-positive streptococci can attach to epithelial cells by a mechanism independent of LTA. When M^+ and M^- streptococci were tested for their binding capacity to human pharyngeal, buccal, and tongue epithelial cells, the M^+ bacteria attached in significantly higher numbers to the pharyngeal cells than to the buccal or tongue cells (Tylewska *et al.*, 1988). The M^- organisms bind in equivalent low numbers to all three oral cell types. While the binding of M^+ streptococci to pharyngeal cells were inhibited by the complete M6 protein molecule as well as LTA, 1,000 fold more LTA than M protein was required to attain equivalent inhibition. Thus, while there may be multiple mechanisms by which streptococci can attach to epithelial cells, it seems apparent that M protein provides the organism with an adherence advantage and consequently, is considered to be an attachment factor.

Non-Type-Specific Protection

The frequency of group A streptococcal respiratory infection rises abruptly at age four, peaks by age six, and declines above age 10, reaching levels commonly found in adults by 18 years (Breese and Hall, 1978). At its peak incidence, 50% of children between the ages of five and seven have streptococcal infections each year. The decreased occurrence of streptococcal pharyngitis in adults might be the consequence of an age-related host factor. Alternatively, protective antibodies directed to antigens common to all group A streptococcal serotypes might arise as a result of multiple infections experienced during childhood, culminating in an elevated response to conserved protective epitopes of the M molecule. While exploring the second hypothesis, it was found that sera of most adults examined have a strong antibody response to the conserved regions of M6 protein (Bessen and Fischetti, 1988c; Fischetti and Windels, 1988). While antibodies directed to conserved epitopes fail to neutralize the antiphagocytic property of M protein (Jones and Fischetti, 1988), they might afford protection by an alternative mechanism, i.e. preventing initial interactions with the mucosa.

The so-called C-repeat region of type 6 streptococci is highly conserved among organisms of many distinct serological types (Jones *et al.*, 1985; Jones *et al.*, 1986; Bessen and Fischetti, 1988a; Bessen *et al.*, 1989; Jones and Fischetti, 1988). Monoclonal antibodies directed to determinants localized to the C-repeat region of the M6 protein bind to the surface of whole streptococci which represent more than half of the serotypes examined (Jones *et al.*, 1985; Jones *et al.*, 1986). In view of the strong evidence for the presence of surface-exposed conserved epitopes on the native M molecule, it should be possible to generate antibodies reactive to these epitopes using for immunization only a few distinct peptides derived from sequences within this region. If a conserved region vaccine is to be effective in preventing nasopharyngeal infection by group A streptococci, it may require stimulation of the secretory immune response.

Experiments employing passive immunization at the mucosa allowed us to more precisely evaluate the role played by secretory IgA (sIgA) in preventing streptococcal infection at a mucosal site (Bessen and Fischetti, 1988b). It was found that affinity-purified human anti-M protein sIgA administered intranasally protected mice against systemic infection after intranasal challenge with group A streptococci. In sharp contrast, anti-M protein specific IgG administered intranasally had no protective effect whatsoever at this site, although it was highly opsonic and promoted streptococcal phagocytosis in whole blood.

An oral or intranasal route of immunization has been used previously for several group A streptococcal vaccines, with the intent of evoking a strong sIgA response (Bessen and Fischetti, 1988a; Poirier *et al.*, 1988; Polly *et al.*, 1975; D'Alessandri *et al.*, 1978; Kurl *et al.*, 1985). To determine whether the conserved, surface exposed epitopes of M protein influence the course of mucosal colonization by group A streptococci, peptides corresponding to these regions were used as immunogens in a mouse mucosal model (Bessen and Fischetti, 1988a). Synthetic peptides corresponding to residues contained within the conserved C-repeat region and an adjacent partially conserved region of the M6 sequence were covalently linked to the mucosal adjuvant cholera toxin B subunit (CTB) and administered intranasally to the mice. About 30 days later, animals were challenged intranasally with live streptococci and pharyngeal colonization monitored by pharyngeal swabs for up to 15 days. It was found that mice immunized with the peptide-CTB complex showed a significant reduction in colonization compared to mice receiving CTB alone. Thus, despite the fact that conserved region peptides are unable to evoke an opsonic antibody response (Jones and Fischetti, 1988), these peptides have the capacity to induce an immune response capable of influencing the colonization of group A streptococci at the nasopharyngeal mucosa in this model system. However, the question still remained as to whether peptides derived from one serotype (i.e., M6) though conserved, could protect against a heterologous serotype in this model.

For these studies a combination of four overlapping synthetic peptides completely contained within the M6 C-repeat (each about 20 amino acids in length) were linked to CTB as in the previous studies and administered to the mice intranasally and orally. The animals this time were challenged intranasally with M14 streptococci and throats were swabbed over a 10-day period. We found that mice immunized with the peptide-CTB complexes displayed significantly lower rates of colonization and death as compared to the control group receiving CTB alone ($p < 0.005$) (Bessen and Fischetti, 1990b). The proportion of mice that were culture positive plus dead on days 1, 2, 3, 6 and 10 were 26%, 35%, 39%, 57%, and 61%, respectively, for the unvaccinated controls and 6%, 9%, 6%, 9%, and 13% for the group that received the peptide-CTB vaccine. These findings suggest for the first time that protection against multiple serotypes of group A streptococci can be achieved with a vaccine consisting of the widely shared C-repeat region of M6 protein.

Vaccinia Virus (VV) as a Vector for Delivering the M Protein Conserved Region

Vaccinia virus has been modified genetically to serve as a vehicle for live recombinant vaccines (Hruby *et al.*, 1988). The successful cloning of the emm6.1 gene into vaccinia virus and its high expression in mammalian cells infected with the recombinant vaccinia may prove to be a powerful vector for delivery of the M molecule to mucosal surfaces (Hruby *et al.*, 1988). In an effort to express only the conserved region epitopes located in the C-terminal half of the molecule, genetic engineering methods were used to remove the 5' half of the M6 gene and recombine the conserved 3' gene fragment into the vaccinia genome (VV:M6'). Western blot analysis revealed that the cells infected with the VV:M6' recombinant produce a molecule of about 30 kDa that is reactive with a monoclonal specific for the C-terminal half of the M6 protein (Fischetti *et al.*, 1989). Mice were immunized intranasally with either the VV:M6' recombinant or wild-type vaccinia and challenged intranasally and orally one month later with 5 x 10^6 streptococci. Pharyngeal cultures taken up to 10 days post challenge revealed that the VV:M6' immunized animals differed significantly from controls.

Of the VV:M6' immunized animals, only 16% of 152 total swabs taken were streptococcal positive with 10 (6%) yielding >100 CFU, whereas

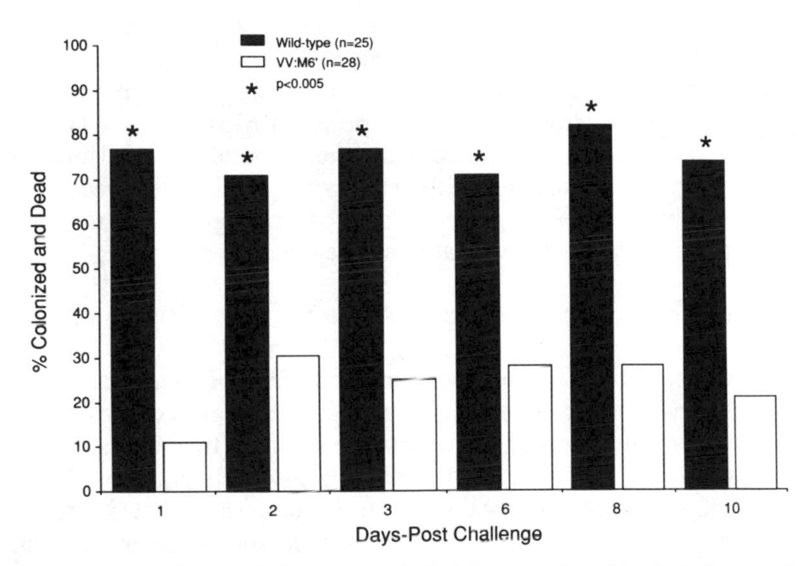

Figure 1. Throat cultures for streptococci after intranasal and oral challenge of mice vaccinated intranasally with wild-type or recombinant vaccinia virus. Pharyngeal cultures were taken up to 10 days following intranasal challenge with 10^6-10^7 streptococci. Results are expressed as the number of culture positive mice plus dead at each time point. Statistical difference was significant according to chi-square analysis.

69% of 115 swabs were positive in the wild-type group and 40 (35%) displaying >100 CFU. On average >70% of the animals immunized with wild-type virus were culture positive for group A streptococci at every swab day up to 10 days after challenge. This is compared with <30% colonization of mice immunized with the VV:M6' recombinant (Figure 1).

In parallel experiments, when VV:M6' vaccinated animals were challenged with heterologous M14 streptococci, protection was also seen when compared to control (Fischetti *et al.*, 1989). These results confirm and extend our previous studies using synthetic peptides from conserved regions of the M molecule to protect against streptococcal colonization (Bessen and Fischetti, 1988a; Bessen and Fischetti, 1990b).

While the vaccinia virus approach has potential for the delivery of bacterial antigens, it is not without concern. For instance, issues pertaining to the safety of vaccinia virus-based vaccines have been raised based on the rare but documented complications resulting from VV immunizations, particularly in immunocompromised individuals. While research involved in the identification of virulence determinants of poxviruses is active, it will be several years before these viruses will be engineered such that they retain their infectivity and immunogenicity without pathogenicity. Vectors such as fowl pox are currently being considered as viable alternatives to the vaccinia virus. These viruses have all the advantages of vaccinia virus without the concerns of side effects. In fact, there have been no recorded human infection as a result of a fowl pox virus. These viruses, which are pathogenic only for avian species, have been used successfully to express the G glycoprotein of rabies virus in non-avian species (Taylor *et al.*, 1988a; Taylor *et al.*, 1988b). Animals immunized with the recombinant fowl pox were protected from lethal challenge with rabies virus. Thus, such vectors may serve as the choice to existing methods for delivering foreign antigens to protect man from bacterial as well as viral diseases.

Conserved Sequence in Gram-Positive Surface Molecules

Proteins on bacterial surfaces perform a variety of functions for the organism from enabling resistance to phagocytosis as seen with the M protein, to adherence to specific receptors on human tissue cells. We have initiated a study to understand the mechanism of attachment of the M molecule within the streptococcal cell in order to devise strategies to block the attachment process. In doing so we may be able to circumvent many problems associated with vaccines directed to surface-exposed epitopes.

The complete sequence of group A streptococcal M6 protein (Hollingshead *et al.*, 1986) has revealed a 20 hydrophobic amino acid segment at the C-terminal end predicted to be a membrane anchor followed by a 6 amino acid charged tail acting as a stop-transfer segment (Blobel, 1980). Sequence analyses of M proteins from other streptococcal serotypes and other surface molecules from gram-positive bacteria (Olsson *et al.*, 1987; Signas *et al.*, 1989; Guss *et al.*, 1984; Kok *et al.*, 1988; Frithz *et al.*, 1989; Ferretti *et al.*, 1989; Okahashi *et al.*, 1989; Schneewind *et al.*, 1990; Haanes and Cleary, 1989; Heath; Cleary, 1989 and Gaillard *et al.*, 1991) have revealed a similar arrangement of hydrophobic amino acids and charged tail despite sequence variation within this region among the M molecules and the other surface proteins. This common motif suggests a conventional mechanism of attachment for these different surface molecules.

```
M6          (375)  PGNKVVPGKGQAPQAGTKPNQNKAPMKETKRQLPSTGETANPFFTAAALTVMATAGVAAVVKRKEEN
wapA        (382)  QTKTTASQTNVPTTTNITTTSKQVTKQKAKFVLPSTGEQAGLLLTTVGLVIVAVAGVYFYRTRR
M49         (283)  ELAKLKGNQTPNAKVAPQANRSRSAMTQQKRTLPSTGETANPFFTAAAATVMVSAGMLALKRKEEN
IgA-BP      (280)  ELAKLKGNQTPNAKVAPQANRSRSAMTQQKRTLPSTGETANPFFTAAAATVMVSAGMLALKRKEEN
Fc-BP       (359)  PDTKPGNKEVPTRPSQTRTNTNKAPMAQTKRQLPSTGEETTNPFFTA...
Protein A   (422)  KLADKNMIKPGQELVVDKKQPANHADANKAQALPETGEENPLIGTTVFGGLSLALGAALLAGRRREL
Protein G   (494)  PIAKDDAKKDDTKKEDAKKPEAKKDDAKKAETLPTTGEGSNPFFTAAALAVMAGAGALAVASKRKED
Fn-BP       (904)  VEQGKVVTPVIEINEKVKAVAPTKKPQSKKSELPETGGEESTNKGMLFGGLFSILGLALLRRNKKNHKA
T6          (472)  KALTDGTTFSKSNEGSGTVLLETDIPNTKLGELPSTGSIGTYLFKAIGSAAMIGAIGIYIVKRRKA
PAc        (1496)  TDPQDPSSPRTSTVIIYKPQSTAYQPSSVQETLPNTGVTNNAYMPLLGIIGLVTSFSLLGLKAKKD
Wg2        (1835)  GGNIPTNPATTTSTSTDDTTDRNGQLTSGKGALPKTGETTERPAFGFLGVIVVILMGVLGLKRKQREE
```

Figure 2. Alignment of C-terminal amino acid sequences from 11 surface proteins from gram-positive cocci. The LPSTGE sequence is shaded and the proteins were aligned along this consensus sequence. In 10 out of 11 proteins, a conserved K residue was found 2 or 3 residues preceeding the consensus sequence (boxed). The C-terminal hydrophobic regions are also boxed. The amino acid number of the first residue of each sequence is in parenthesis. Abbreviations: M6: M protein (*S. pyogenes*) (Hollingshead *et al.*, 1986); IgA-BP: IgA binding protein (from an M4 *S. pyogenes*) (Frithz *et al.*, 1989); wapA: wall-associated protein A (*S. mutans*) (Ferretti *et al.*, 1989); Fc-BP: Fc binding protein from *S. pyogenes* (Heath and Cleary, 1989); Protein G: IgG binding protein (group G streptococi) (Olsson *et al.*, 1987); PAc: cell surface protein (*S. mutans*) (Okahashi *et al.*, 1989); Protein A: IgG binding protein (*S. aureus*) (Guss *et al.*, 1984); wg2: cell wall protease (*S. cremoris*) (Kok *et al.*, 1988); Fn-BP: fibronectin binding protein (*S. aureus*) (Signas *et al.*, 1989); T6: surface protein (*S. pyogenes*) (Schneewind *et al.*, 1990).

Examination of the C-terminal region of surface molecules from gram-positive cocci has revealed a hexapeptide sequence which begins 9 residues N-terminal to the putative membrane anchor region. This sequence is found in virtually all sequenced surface molecules from gram-positive cocci (particularly streptococcal and staphylococcal species) (Fischetti *et al.*, 1990) (Figure 2). The hexapeptide, with the consensus sequence LPSTGE, has recently (Gaillard *et al.*, 1991) been reported in a protein from gram-positive bacilli, but is not as yet been found in gram-negative organisms. These results suggest that a common motif may be used for attaching these surface proteins in the gram-positive cell. This discovery may open new avenues for controlling infection by these organisms. Because surface proteins are used by disease bacteria to either initiate or maintain infection, by devising methods to prevent these molecules from anchoring within the bacterial cell, we should be able to block infection and thus, circumvent some of the problems associated with antibiotic therapy.

The proximity of the LPSTGE sequence to the hydrophobic amino acid segment of these bacterial proteins is reminiscent of a similarly located hexapeptide sequence in some glycosyl-phosphatidylinositol (GPI)-anchored molecules (Ferguson and Williams, 1988). Although a GPI-like anchor complex has not been as yet identified in procaryotes, our evidence indicates that streptococcal M protein is anchored directly to the cytoplasmic membrane and is found both as a bound and released form (Pancholi and Fischetti, 1989). Thus, the LPSTGE motif is a finding which may have significant implications regarding the attachment of surface proteins in gram-positive organisms.

References

Beachey, E.H., Simpson, A., Ofek, I., Hasty, D.L., Dale, J.B. and Whitnack, E., 1983, Attachment of *Streptococcus pyogenes* to mammalian cells, *Rev. Infect. Dis.* 5:S670.

Bessen, D. and Fischetti, V.A., 1988a, Influence of intranasal immunization with synthetic peptides corresponding to conserved epitopes of M protein on mucosal colonization by group A streptococci, *Infect. Immun.* 56:2666.

Bessen, D. and Fischetti, V.A., 1988b, Passive acquired mucosal immunity to group A streptococci by secretory immunoglobulin A, *J. Exp. Med.* 167:1945.

Bessen, D. and Fischetti, V.A. 1988c, Vaccination against *Streptococcus pyogenes* infection, in: New Generation Vaccines, Levine, M.M. and Wood, G., Eds., Marcel Dekker,Inc., New York. p 599.

Bessen, D. and Fischetti, V.A. 1988d, Role of nonopsonic antibody in protection against group A streptococcal infection, in: Technological Advances in Vaccine Development, Lasky, L., Ed., Alan R. Liss Inc., New York. p 493.

Bessen, D. and Fischetti, V.A., 1990a, Differentiation between two biologically distinct classes of group A streptococci by limited substitutions of amino acids within the shared region of M protein-like molecules, *J. Exp. Med.* 172: 1757.

Bessen, D. and Fischetti, V.A., 1990b, Synthetic peptide vaccine against mucosal colonization by streptococci: Protection against a heterologous M type with a conserved C-repeat region epitope, *J. Immunol.* 145:1251.

Bessen, D., Jones, K.F. and Fischetti, V.A., 1989, Evidence for two distinct classes of streptococcal M protein and their relationship to rheumatic fever, *J. Exp. Med.* 169:269.

Blobel, G., 1980, Intracellular protein topogenesis, *Proc. Natl. Acad. Sci. U.S.A.* 77:1496.

Breese, B.B. and Hall, C.B. (1978) Beta hemolytic streptococcal diseases. Houghton Mifflin, Boston.

D'Alessandri, R., Plotkin, G., Kluge, R.M., Wittner, M.K., Fox, E.N., Dorfman, A. and Waldman, R.H., 1978, Protective studies with group A streptococcal M protein vaccine. III. Challenge of volunteers after systemic or intranasal immunization with type 3 or type 12 group A Streptococcus, *J. Infect. Dis.* 138:712.

Ellen, R.P. and Gibbons, R.J., 1972, M protein-associated adherence of *Streptococcus pyogenes* to epithelial surfaces: prerequisite for virulence, *Infect. Immun.* 5:826.

Ferguson, M.A.J. and Williams, A.F., 1988, Cell-surface anchoring of proteins via glycosyl-phosphatidylinositol structures, *Ann. Rev. Biochem.* 57:285.

Ferretti, J.J., Russell, R.R.B. and Dao, M.L., 1989, Sequence analysis of the wall-associated protein precursor of *Streptococcus mutans* antigen A, *Molec. Microbiol.* 3:469.

Fischetti, V.A., Hodges, W.M. and Hruby, D.E., 1989, Protection against streptococcal pharyngeal colonization with a vaccinia:M protein recombinant, *Science* 244:1487.

Fischetti, V.A., Pancholi, V. and Schneewind, O., 1990, Conservation of a hexapeptide sequence in the anchor region of surface proteins of gram-positive cocci, *Molec. Microbiol.* 4:1603.

Fischetti, V.A. and Windels, M., 1988, Mapping the immunodeterminants of the complete streptococcal M6 protein molecule: Identification of an immunodominant region, *J. Immunol.* 141:3592.

Flores, A.E., Johnson, D.R., Kaplan, E.L. and Wannamaker, L.W., 1983, Factors influencing antibody responses to streptococcal M proteins in humans, *J. Infect. Dis.* 147:1.

Frithz, E., Heden, L-O. and Lindahl, G., 1989, Extensive sequence homology between IgA receptor and M protein in *Streptococcus pyogenes*, *Molec. Microbiol.* 3:1111.

Gaillard, J.L., P. Berche, C. Frehel, E. Gouin and P. Cossart, 1991, Entry of *L. monocytogenes* into cells mediated by a repeat protein analogous to surface antigens from gram-positive, extracellular pathogens, *Cell*, 65: 1-20.

Guss, B., Uhlen, M., Nilsson, B., Lindberg, M., Sjoquist, J. and Sjodahl, J., 1984, Region X, the cell-wall-attachment part of staphylococcal protein A, *Eur. J. Biochem.* 138:413.

Haanes, E.J. and Cleary, P.P., 1989, Identification of a divergent M protein gene and an M protein related gene family in serotype 49 *Streptococcus pyogenes*, *J. Bacteriol.* 171:6397.

Heath, D.G. and Cleary, P.P., 1989, Fc-receptor and M protein genes of group A streptococci are products of gene duplication, *Proc. Natl. Acad. Sci. U.S.A.* 86:4741.

Hollingshead, S.K., Fischetti, V.A. and Scott, J.R., 1986, Complete nucleotide sequence of type 6 M protein of the group A streptococcus: repetitive structure and membrane anchor, *J. Biol. Chem.* 261:1677.

Horstmann, R.D., Sievertsen, H.J., Knobloch, J. and Fischetti, V.A., 1988, Antiphagocytic activity of streptococcal M protein: selective binding of complement control protein Factor H, *Proc. Natl. Acad. Sci. USA* 85:1657.

Hruby, D.E., Hodges, W.M., Wilson, E.M., Franke, C.A. and Fischetti, V.A., 1988, Expression of streptococcal M protein in mammalian cells, *Proc. Natl. Acad. Sci. USA* 85:5714.

Jones, K.F. and Fischetti, V.A., 1988, The importance of the location of antibody binding on the M6 protein for opsonization and phagocytosis of group A M6 streptococci, *J. Exp. Med.* 167:1114.

Jones, K.F., Khan, S.A., Erickson, B.W., Hollingshead, S.K., Scott, J.R. and Fischetti, V.A., 1986, Immunochemical localization and amino acid sequence of cross-reactive epitopes within the group A streptococcal M6 protein, *J. Exp. Med.* 164:1226.

Jones, K.F., Manjula, B.N., Johnston, K.H., Hollingshead, S.K., Scott, J.R. and Fischetti, V.A., 1985, Location of variable and conserved epitopes among the multiple serotypes of streptococcal M protein, *J. Exp. Med.* 161:623.

Kok, J., Leenhouts, K.J., Haandrikman, A.J., Ledeboer, A.M. and Venema, G., 1988, Nucleotide sequence of the cell wall proteinase gene of *Streptococcus cremoris* Wg-2, *Appl. Environ. Microbiol.* 54:231.

Kurl, D.N., Stjernquist-Desatnik, A., Schalen, C. and Christensen, P., 1985, Induction of local immunity to group A streptococci type M50 in mice by non-type-specific mechanisms, *Acta Path. Microbiol. Immunol. Scand. Sect. B* 93:401.

Lancefield, R.C., 1928, The antigenic complex of *Streptococcus hemolyticus*. I Demonstration of a type-specific substance in extracts of *Streptococcus hemolyticus*, *J. Exp. Med.* 47:91.

Lancefield, R.C., 1962, Current knowledge of the type specific M antigens of group A streptococci, *J. Immunol.* 89:307.

Little, C. 1981, Phospholipase C from *Bacillus cereus*, in: Methods in Enzymology, Academic Press, New York. p 725.

Nagel, S.D. and Boothroyd, J.C., 1989, The major surface antigen, P30, of *Toxoplasma gondii* is anchored by a glycolipid, *J. Biol. Chem.* 264:5569.

Okahashi, N., Sasakawa, C., Yoshikawa, S., Hamada, S. and Koga, T., 1989, Molecular characterization of a surface protein antigen gene from serotype c *Streptococcus mutans* implicated in dental caries, *Molec. Microbiol.* 3:673.

Olsson, A., Eliasson, M., Guss, B., Nilsson, B., Hellman, U., Lindberg, M. and Uhlen, M., 1987, Structure and evolution of the repetitive gene encoding streptococcal protein G, *Eur. J. Biochem.* 168:319.

Pancholi, V. and Fischetti, V.A., 1989, Identification of an endogeneous membrane anchor-cleaving enzyme for group A streptococcal M protein, *J. Exp. Med.* 170:2119.

Phillips, G.N., Flicker, P.F., Cohen, C., Manjula, B.N. and Fischetti, V.A., 1981, Streptococcal M protein: α-helical coiled-coil structure and arrangement on the cell surface, *Proc. Natl. Acad. Sci. USA.* 78:4689.

Poirier, T.P., Kehoe, M.A. and Beachey, E.H., 1988, Humoral and mucosal immunity against group A streptococci evoked by attenuated aroA *Salmonella typhimurium* expressing cloned streptococcal M protein, *J. Exp. Med.* 168:25.

Polly, S.M., Waldman, R.H., High, P., Wittner, M.K., Dorfman, A. and Fox, E.N., 1975, Protective studies with a group A streptococcal M protein vaccine. II. Challenge of volunteers after local immunization in the upper respiratory tract, *J. Infect. Dis.* 131:217.

Robbins, J.C., Spanier, J.G., Jones, S.J., Simpson, W.J. and Cleary, P.P., 1987, *Streptococcus pyogenes* type 12 M protein regulation by upstream sequences, *J. Bacteriol.* 169:5633.

Schneewind, O., Jones, K.F. and Fischetti, V.A., 1990, Sequence and structural characterization of the trypsin-resistant T6 surface protein of group A streptococci, *J. Bacteriol.* 172:3310.

Scott, J.R., Hollingshead, S.K. and Fischetti, V.A., 1986, Homologous regions within M protein genes in group A streptococci of different serotypes, *Infect. Immun.* 52:609.

Signas, C., Raucci, G., Jonsson, K., Lindgren, P., Anantharamaiah, G.M., Hook, M. and Lindberg, M., 1989, Nucleotide sequence of the gene for fibronectin-binding protein from *Staphylococcus aureus*: Use of this peptide sequence in the synthesis of biologically active peptides, *Proc. Natl. Acad. Sci. USA* 86:699.

Taylor, J., Weinberg, R., Kawoaka, Y., Webster, R.G. and Paoletti, E., 1988a, Protective immunity against avian influenza induced by a fowlpox virus recombinant, *Vaccine* 6:504.

Taylor, J., Weinberg, R., Languet, B., Desmettre, P. and Paoletti, E., 1988b, Recombinant fowlpox virus inducing protective immunity in non-avian species, *Vaccine* 6:497.

Tylewska, S.K., Fischetti, V.A. and Gibbons, R.J., 1988, Binding selectivity of *Streptococcus pyogenes* and M-protein to epithelial cells differs from that of lipoteichoic acid, *Curr. Microbiol.* 16:209.

Group A Streptococci: Molecular Mimicry, Autoimmunity and Infection

Madeleine W. Cunningham

Introduction

Group A streptococci produce a number of suppurative infections in man including acute pharyngitis, impetigo, cellulitis, erysipelas and septicemia (Bisno, 1979a). Two sequelae, acute rheumatic fever or acute glomerulonephritis, may follow infection and may be related to hyper-immune responsiveness in the host (Bisno, 1979 b,c). A number of reports have suggested that the post-streptococcal sequelae are a result of molecular mimicry between streptococcal M proteins and α-helical coiled-coil proteins in the host tissues (Krisher and Cunningham, 1985; Dale and Beachey, 1985a, 1986; Manjula and Fischetti, 1986; Cunningham and Swerlick, 1986; Cunningham *et al.*, 1989; Fenderson *et al.*, 1989; Kraus *et al.*, 1989). Streptococcal membrane proteins have also been implicated in the immunological crossreactions between streptococci and heart (Zabriskie and Freimer, 1966; van de Rijn *et al.* 1977; Cunningham *et al.*, 1984; Barnett and Cunningham, 1990).

Heart-reactive antibodies were first found in the sera of patients with acute rheumatic fever (Zabriskie *et al*, 1980). Furthermore, streptococcal wall-membrane antigens were found to induce anti-heart antibodies in animals (Zabriskie and Freimer, 1966; van de Rijn *et al.*, 1980; Ayakawa *et al.*, 1988), and the membranes were found to absorb the heart-reactive antibodies from the sera. The presence of anti-heart antibodies in acute rheumatic fever sera and the deposition of immunoglobulin and complement in heart tissue of acute rheumatic fever patients supports the hypothesis that acute rheumatic fever has an autoimmune origin (Zabriskie *et al.*, 1970; Kaplan, 1963). If acute rheumatic fever is an autoimmune disease as proposed, it is an important model for study of the role of infectious agents in the development of autoimmunity in man. A recent review addresses the role of the streptococcus in molecular mimicry and autoimmunity (Froude *et al.*, 1989).

Streptococcal components most recently implicated in immunological crossreactions with heart tissue include streptococcal membrane proteins (van de Rijn *et al.*, 1977; Barnett and Cunningham, 1990) and M proteins, such as types 1, 5, 6, 12 and 19 (Dale and Beachey, 1985 a,b, 1986;

Cunningham and Swerlick 1986; Bronze *et al.*, 1988; Cunningham *et al.*, 1989; Fenderson *et al.*, 1989; Kraus *et al.*, 1989). Since M protein is anti-phagocytic and is the major virulence factor of group A streptococci, protection of the host is dependent on the immune response against the M protein antigen (Lancefield, 1962; Bessen *et al.*, 1989; Fischetti *et al.*, 1989). The vigorous hyperresponsiveness observed in acute rheumatic fever to M protein, membrane proteins, carbohydrate groups and other streptococcal antigens (Read *et al.*, 1974; Reid *et al.*, 1980; Barnett and Cunningham, 1990; Shulman and Ayoub, 1974) may be responsible for heart-reactive antibodies present in sera of rheumatic fever patients. Increased incidence of rheumatic fever in certain HLA phenotypes has been reported (Ayoub *et al.*, 1986). Although a specific genetic defect has never been found in streptococcal sequelae, a specific marker for acute rheumatic fever was found to be present on peripheral blood B lymphocytes (Zabriskie *et al.*, 1987), and a specific anti-myosin idiotype was present among antibodies in acute rheumatic fever sera (McCormack, *et al.*, submitted).

Host tissue components crossreactive with streptococcal M protein and membrane proteins are clearly α-helical coiled-coil molecules such as myosin, tropomyosin, keratin, vimentin and laminin (Cunningham *et al.*, 1985a; Dale and Beachey, 1985, 1986; Cunningham and Swerlick, 1986; Fenderson *et al.*, 1989; Kraus *et al.*, 1989; Antone and Cunningham, in preparation). Structural analysis of M proteins has led to the conclusion that they are α-helical coiled-coil structures with a seven amino acid residue periodicity (Fischetti and Manjula, 1982). M proteins were compared to tropomyosin, myosin and the desmin-keratin family of α-helical proteins (Manjula, *et al.*, 1985). Crossreactive epitopes of M proteins, myosin and vimentin have been defined (Dale and Beachey, 1986; Bronze *et al*, 1988; Cunningham *et al.*, 1989; Kraus *et al.*, 1989; Dell *et al.*, 1991) and may be responsible for the immunological crossreactivity observed between streptococci and heart and other tissues.

Crossreactive α-helical molecules of the host are found in tissues which may become inflamed during acute rheumatic fever, namely, the heart, joint, skin or brain. Major clinical manifestations of acute rheumatic fever include carditis, arthritis, subcutaneous nodules, erythema marginatum or chorea (Stollerman, 1988). The data would suggest that the crossreactive immune responses result in inflammation in these respective tissues.

Crossreactive Monoclonal Antibody Probes

In our studies we have obtained evidence of immunological crossreactivity between anti-streptococcal monoclonal antibodies and myosin (Krisher and Cunningham, 1985). These antibodies were first shown to react with whole streptococci and membranes but were also found to react with M proteins (Cunningham and Russell, 1983; Cunningham *et al.*, 1984; Cunningham and Swerlick, 1986; Fenderson *et al.*, 1989; Cunningham *et al.*, 1989).

Table I

Antigen binding of crossreactive monoclonal antibody probes

mAb	Antigen Specificity
6.5.1	Actin/M proteins 5,6
8.5.1	Myosin/Streptococci
24.1.2	Myosin/M proteins 5,6
27.4.1	Myosin/Tropomyosin/Keratin/M protein 5,6
36.2.2	Myosin/Tropomyosin/Keratin/Actin/ Laminin/Mproteins 1,5,6
40.4.1	Myosin/M Protein 5
49.8.9	Vimentin/M protein 5,6
54.2.8	Myosin/Tropomyosin/Vimentin/DNA/ M proteins 5,6
55.4.1	DNA/M protein 5,6
101.4.1	Myosin/Streptococci
112.2.2	Myosin/Tropomyosin/M proteins 5,6
113.2.1	Actin/M proteins 5, 6
654.1.1	Myosin/Tropomyosin/DNA/M proteins 5,6

Thirteen monoclonal antibody probes studied over the past several years were produced following immunization of BALBc/BYJ or BALBc mice with M type 5 membranes, Pep M5 protein, or mutanolysin extracted M protein (Cunningham and Russell, 1983; Fenderson *et al.*, 1989). The hybridomas were screened for antibody production in the enzyme linked immunosorbent assay (ELISA) using whole streptococci and human heart extracts as antigens (Cunningham and Russell, 1983). The monoclonal antibodies (Table I) selected for study were of the IgM isotype (Cunningham *et al.*, 1984). The specificities of these thirteen crossreactive monoclonal antibody probes are reviewed in Table I.

Many of the monoclonal antibody (mAb) probes recognized myosin and tropomyosin and reacted with tissue sections of human heart as shown in Figure 1. Both sarcolemmal and subsarcolemmal components reacted with mAbs to various degrees reflecting the specificities described in Table I. Figure 2 demonstrates the anti-nuclear antibody reactivity of mAb 654.1.1. Similar anti-nuclear reactivity has been shown for mAbs 54.2.8 (Cunningham and Swerlick, 1986) and 55.4.1 (data not shown).

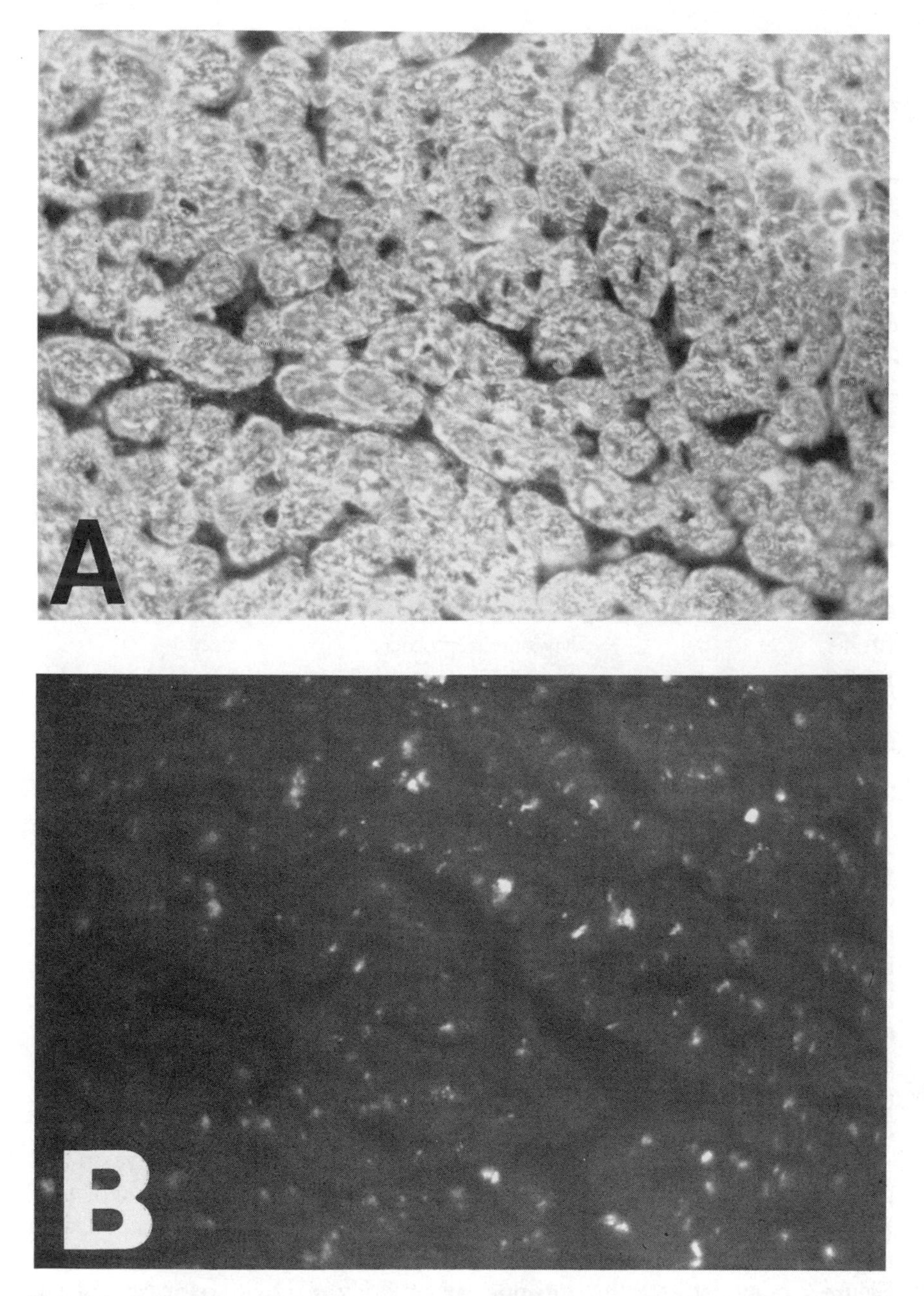

Figure 1. Indirect immunofluorescence of crossreactive mouse mAb 112.2.2 (10ug/ml IgM) labeling a section of human ventricle (A) and the control which was reacted with mouse IgM, (20 ug/ml) (B).

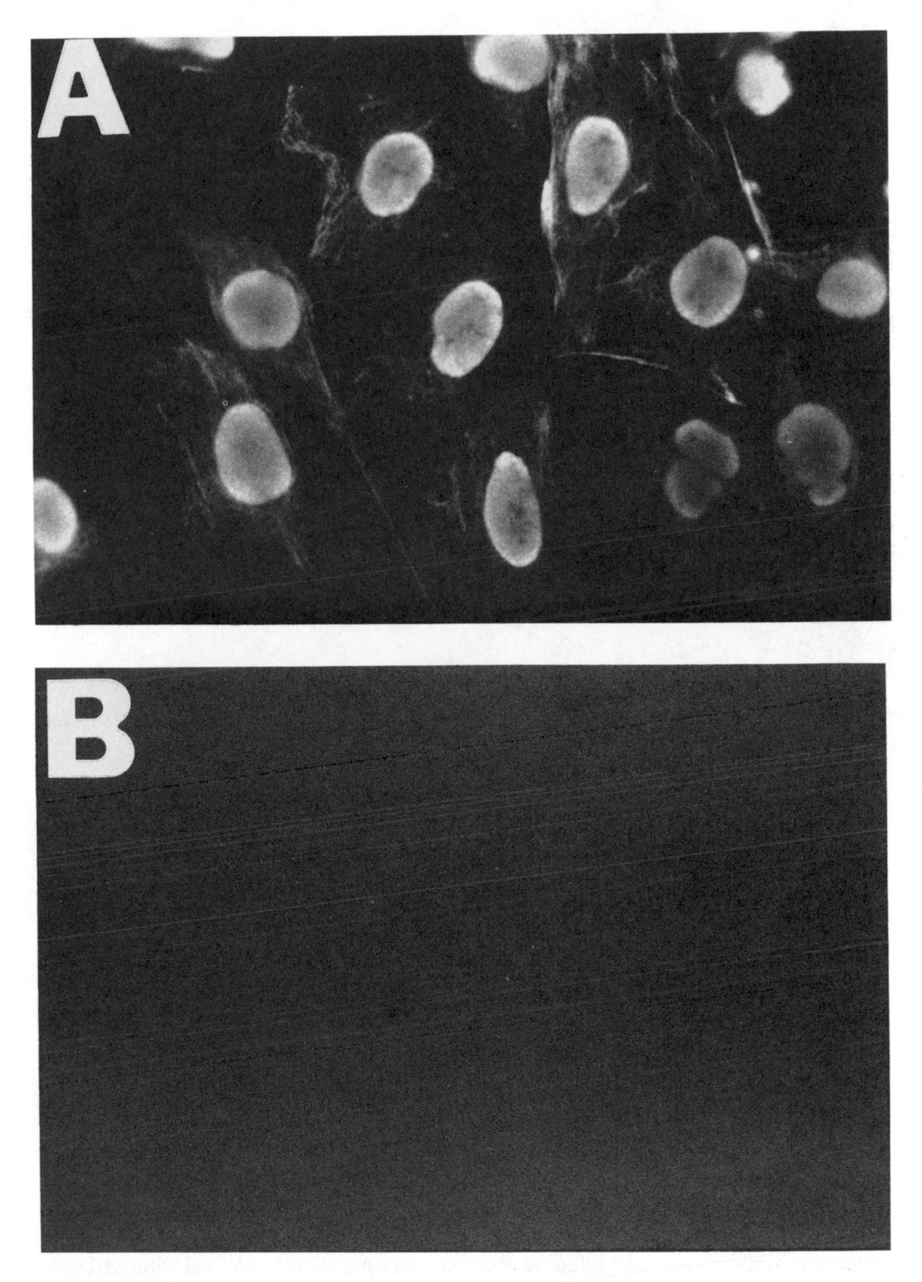

Figure 2. Indirect immunofluorescent labeling of rat heart cell nuclei by mAb 654.1.1 (5 ug/ml IgM) but not by mouse IgM (20 ug/ml) (B). Cells were acetone-fixed prior to reaction with antibody.

The entire mAb panel reacted strongly with whole group A streptococci in ELISA, including 10 different M protein serotypes (unpublished data). Only types 1, 5 and 6 have been investigated thoroughly for reactivity with the mAbs. mAb 36.2.2 was the only mAb probe to recognize all three M proteins while many of the mAbs recognized both M proteins 5 and 6. A crossreactive epitope of types 5 and 6 has been defined and includes a pentameric amino acid sequence (QKSKQ) identical in M5 and M6 proteins (Cunningham *et al.*, 1989). The epitope was located in the B repeat region near the pepsin cleavage site of the M protein (Fischetti *et al.*, 1988). Synthetic peptides of M5 protein containing amino acid residues 164-197 or 184-197 were recognized by mAbs 6.5.1, 27.4.1, 654.1.1, 36.2.2, and 54.2.8 but not by any other mAb probes listed in Table I. Human monoclonal antibody probes, as well as anti-myosin antibodies affinity purified from acute rheumatic fever serum, were reactive with the same M5 epitope (Table II) (Cunningham *et al.*, 1989). Table II illustrates a competitive inhibition of affinity-purified anti-myosin antibody from two acute rheumatic fever and rheumatic heart disease sera by synthetic peptides of M5 protein.

Table II

Competitive inhibition of affinity-purified myosin antibody from two acute rheumatic fever and two rheumatic heart disease sera with selected peptides of PepM5 protein in ELISA

	Antibodies			
Synthetic Peptide[a]	1	2	3	4
28-54	0[a]	0	0	0
84-116	71	0	0	0
164-197	0	0	0	0
184-197	68	87	71	84
188-197	0	0	0	0
PBS	0	0	0	0
PepM5	84	92	85	90

[a]Percent inhibition was calculated on the basis of an antibody control diluted 1/2 in PBS and reacted with PepM5. (Reprinted with permission from the *J. Immunol.*)

These studies suggest that M proteins 5 and 6 contain an epitope (residues 184-188 of M5) that reacts with anti-myosin antibodies from acute rheumatic fever or heart disease. A secondary epitope was located in the 84-116 amino acid sequence of M5 protein (Table II, #2).

Immunologic crossreactivity between M5 and 6 proteins and tropomyosin by mAbs 27.4.1, 36.2.2, 54.2.8, 112.2.2 and 654.1.1 has been demonstrated (Fenderson *et al.*, 1989). Significant sequence homology between the M6 protein and human cardiac tropomyosin was found in the carboxy terminal C-repeat region of M protein. The amino acid residues in this region are more conserved among the M proteins than the residues found in the amino terminal region (Hollingshead *et al.*, 1987).

Localization of Crossreacting Epitopes in Human Cardiac Myosin and Other α-Helical Coiled-Coil Molecules

Immunological crossreactions between M protein and human cardiac myosin have been investigated and the epitopes in cardiac myosin defined (Figure 3) (Dell *et al.*, 1991). Proteolytic fragments of human cardiac

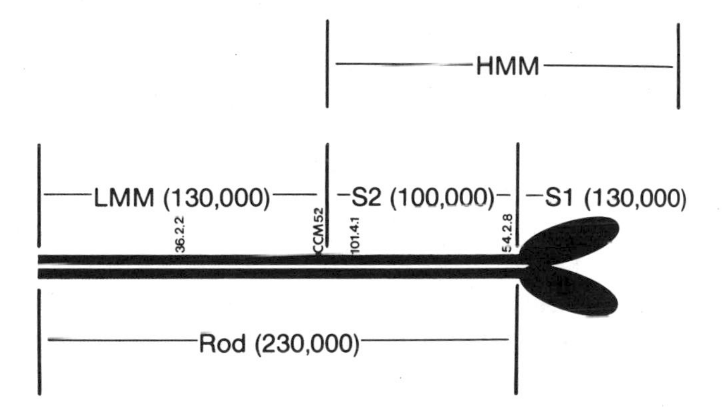

Figure 3. Diagram of myosin heavy chain and its subfragments with the estimated locations of the crossreactive epitopes placed along the rod segment.

myosin were used to identify sites of crossreactivity in the α-helical rod of the myosin heavy chain. None of the mAb panel (Table I) reacted with the S1 or the head fragment of human cardiac myosin. All of the myosin epitopes crossreactive with M protein were located in the α-helical rod. MAb 36.2.2 and 54.2.8 reacted with a site in the light meromyosin subfragment (LMM) while 54.2.8 also recognized an epitope in the S2 subfragment of the rod (Dell *et al.*, 1991). An epitope shared between M protein and myosin in the LMM fragment has been localized to a 14 amino acid fragment which shows approximately 40% homology with the M6 protein (Cunningham *et al.*, 1992). The homologous M protein sequence was located in the C-repeat region in the same segment that shared significant homology with human cardiac tropomyosin (Fenderson *et al.*, 1989).

As shown in Table I, mAb 101.1.4 crossreacted with group A streptococci and myosin and reacted with an epitope in the S2 subfragment of human cardiac myosin (Figure 3). MAb CCM-52, originally produced to rabbit cardiac myosin reacted with the rod and has been reported to recognize light meromyosin (Clark *et al.*, 1984; Eisenberg *et al.*, 1987). MAb CCM-52 also reacted with streptococcal M proteins 5 and 6 (Barnett and Cunningham, submitted).

Differences in the specificity of the mAb panel for skeletal and cardiac myosins was subtle with preferences for cardiac or skeletal myosin (unpublished data). Studies of rabbit and human tropomyosins also suggested subtle differences in mAb recognition of antigen. MAb 27.4.1 reacted with rabbit cardiac tropomyosin but not with human cardiac tropomyosin in Western blots, and mAb 112.2.2 recognized only human cardiac tropomyosin. Only one amino acid difference has been observed between rabbit and human tropomyosin at residue 220 (Mische *et al.*, 1987).

Antigens recognized by heart crossreactive antibodies appeared to be molecules containing α−helical structures such as myosin, tropomyosin, keratin, vimentin, and laminin (Krisher and Cunningham, 1985, Fenderson *et al.*, 1989; Cunningham and Swerlick, 1986; Kraus *et al.*, 1989 and Antone and Cunningham, in preparation). The mAb panel did not recognize bovine serum albumin, Ro antigen, lysozyme, IgG, collagen and histones, indicating that crossreactions may be primarily limited to α−helical molecules. Several crossreactive mAbs recognized actin which is not an α−helical molecule but does polymerize to form filaments or F-actin. Further studies will be required to determine the limits of the immunological crossreactivity of these mAbs.

Our most recent studies show that myosin and vimentin are crossreactive molecules in human heart valves (Gulizia *et al.*, 1991). Purified vimentin inhibited the reaction of mAbs 49.8.9 and 54.2.8 with valve sections. Valvular interstitial cells were highly reactive with mAbs and proved to contain large amounts of vimentin. Crossreactive epitopes of M protein and vimentin recognized by our mAbs have not yet been defined. Valvular interstitial cells might be responsible for collagen deposition in inflamed valves during and following episodes of acute rheumatic fever (Gulizia *et al.*, 1991). In addition, mAbs 36.2.2, 49.8.9 and 54.2.8 labeled valvular surface endothelium, elastin fibrils, and myocytes as well as valvular interstitial cells. If these mAbs react with the valve in vivo, they could trigger inflammation which then might lead to collagen deposition on the valve and subsequent valve dysfunction (Gulizia *et al.*, 1991).

Search for Actin-Like and Myosin-Like Proteins in Group A Streptococci

It was reported previously that streptococcal membranes contained antigens crossreactive with group A streptococci and heart (Zabriskie and Freimer, 1966 and van de Rijn *et al.*, 1977). However, the antigens associated with this crossreactivity were never completely defined. In order to identify other crossreactive streptococcal antigens, separate from M protein, mAbs were used to study extracts from M+ and M- group A streptococci (Barnett and Cunningham, 1990). Figure 4 shows a Western blot of an extract from an M type 5 strain of *Streptococcus pyogenes*, reacted with mAbs 36.2.2 and 54.2.8. The mAbs reacted with a 60 kDa protein not recognized by mAb 10B6, which is an antibody specific for M proteins. The 60 kDa antigen was present also in an M negative strain (T28/51/4) which had spontaneously lost

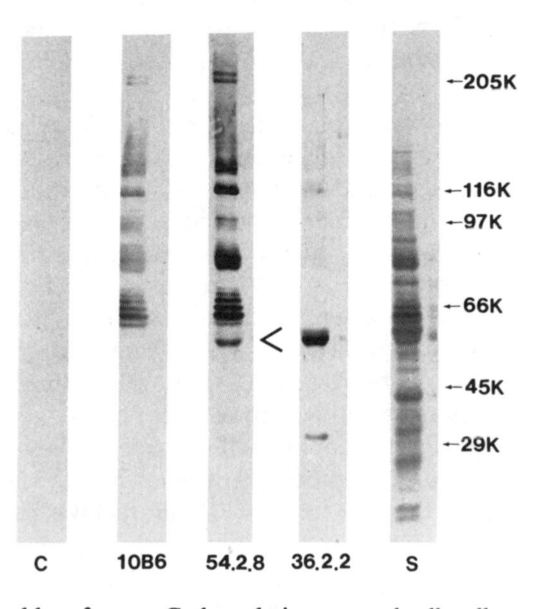

Figure 4. Western blot of group C phage lysin-extracted cell wall-associated proteins of strain M type 5 *Streptococcus pyogenes* reacted with M protein-specific mAb 10B6 and heart crossreactive mAbs 54.2.8 and 36.2.2; 54.2.8 and 36.2.2 reacted with 60-kDa protein (arrowhead). S, Amido black protein stain of the streptococcal extract; C, medium control containing Iscove's modified Dulbecco's medium (GIBCO) plus 10% horse serum. Molecular weight standard proteins are indicated at right (in thousands). (Reprinted with permission from the *J. Inf. Dis.*)

its gene for type 28 M protein. Streptococcal membranes also contained the 60 kDa antigen recognized by mAbs 36.2.2, 54.2.8, and anti-myosin antibodies affinity purified from acute rheumatic fever sera (Barnett and Cunningham, 1990). Since mAbs 36.2.2 and 54.2.8 reacted with the 60 kDa protein, it must share epitopes with M protein as well as with actin and myosin.

It is not known if streptococci contain actins or myosins. However, previous studies demonstrated that M protein formed filamentous arrays when mixed with eukaryotic actin (Chalovich and Fishetti, 1986). M protein does not appear to be a myosin since it does not demonstrate ATPase activity (unpublished data). In our investigations, we reacted anti-actin (NEI-051, DuPont) and anti-myosin (gift of Dr. William Clark, Northwestern University, Chicago, IL) mAbs with extracts from two M+ and one M- *S. pyogenes* strain. Anti-myosin mAb CCM-52 reacted with a protein of approximately 70 kDa in the M- extract and M proteins in the extracts from M+ strains (Barnett and Cunningham, in press). Anti-actin mAb reacted strongly with proteins from 45kDa to 70 kDa. It is not clear if these mAbs recognized true actins and myosins since drosophila actin and rabbit cardiac myosin gene probes did not hybridize with streptococcal DNA (Barnett and Cunningham, unpublished data). These data indicated that nucleotide sequences encoding eukaryotic actin or myosin were not present in streptococci.

Nevertheless, actin was extracted from group A, M negative streptococci with acetone and polymerized under conditions used to polymerize eukaryotic actin (Clarke and Spudich, 1974). Actin-like streptococcal protein obtained by this procedure yielded a large amount of a 60 kDa protein with minor proteins of higher and lower molecular size. Bacterial actin inhibited binding of an anti-actin mAb as well as eukaryotic actin in indirect immunofluorescence tests. Bacterial actin bound strongly to DNAse I and also activated myosin ATPase activity of human cardiac myosin, which are characteristics of eukaryotic actin (Barnett and Cunningham, in press). Filaments of the bacterial actin-like protein were observed under conditions where actin filaments form para-crystalline arrays (Moore *et al.*, 1970). Properties of the actin-like streptococcal protein are listed in Table III.

In further studies, the gene for a 60 kDa protein was cloned using acute rheumatic fever sera (Barnett *et al.*, in preparation). Recombinant protein reacted with affinity purified anti-myosin and anti-actin antibodies from acute rheumatic fever sera. The nature of the 60 kDa protein is under investigation. It shares epitopes with M protein since the affinity purified anti-myosin antibodies from acute rheumatic fever sera crossreacted with M protein. It will be interesting to determine the properties of the 60 kDa recombinant protein and the nucleotide sequence of its gene. Most importantly, antibodies to myosin in acute rheumatic fever react strongly with the 60 kDa protein.

Table III

Properties of an actin-like streptococcal protein

Activates Myosin ATPase Activity
Binds to DNAse I
Makes filaments in presence of ATP, Mg++, K+
Inhibits the binding of anti-actin mAb

Cytotoxicity Linked with Crossreactive mAb

Can autoantibodies produced in response to streptococcal infection play a role in the pathogenesis of acute rheumatic fever and rheumatic carditis? To determine if any of our mAb panel were cytotoxic to heart cells, the antibodies were reacted with primary rat heart cells in the presence of complement (Cunningham *et al.*, 1992, Antone and Cunningham, in preparation). MAb 36.2.2 was the most cytolytic antibody of the panel producing detectable lysis at 1 ug/ml of antibody. Ten times as much antibody was required to observe lysis with mAbs 49.8.9 and 54.2.8. Rat fibroblasts were also sensitive targets, however, a rat liver cell line was not (Antone and Cunningham, in preparation).

MAb 36.2.2 labeled the surface of unfixed heart cells in culture (figure 5). None of the other mAbs in the panel produced such strong reactions with the heart cell surface or were as cytotoxic. MAb 36.2.2 was the only mAb of the panel (Table I) which reacted with purified laminin. Further experiments showed that soluble laminin or rat heart cells inhibited the reaction of mAb 36.2.2 with laminin in Western blots. Cross-linking of mAb 36.2.2 to heart cell surface proteins has not been successful leaving the identification of the heart cell surface protein equivocal. However, the data strongly suggest that laminin, with its α-helical coiled-coil domains of A, B1 and B2 chains, may be the heart cell surface protein recognized by mAb 36.2.2. Significant amino acid homology between M protein and laminin is found within the α-helical coiled-coil segments of laminin. Until experiments permit further analysis of the mAb reaction, the evidence will remain circumstantial that laminin is the target antigen responsible for cytolysis by mAb 36.2.2. The presence of mAbs like 36.2.2 in human sera might be damaging to the host and play a role in development of carditis.

Most recent data show that M5 protein amino acid sequences are crossreactive with laminin (Antone and Cunningham, in preparation).

Overlapping synthetic peptides (18-mers) containing M5 protein amino acid sequences were linked to keyhole limpet hemocyanin and used to immunize Balb/c mice. Antibodies to two specific peptides reacted with laminin (Antone and Cunningham, in preparation).

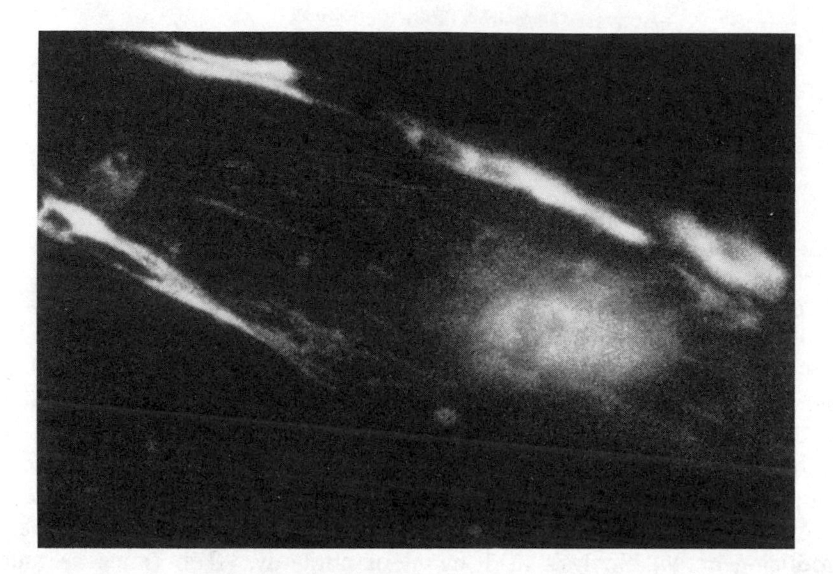

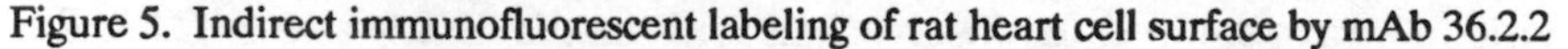

Figure 5. Indirect immunofluorescent labeling of rat heart cell surface by mAb 36.2.2

Lymphocytes (primarily T cells) from lymph nodes of Balb/c mice immunized with laminin were tested for responsiveness to the M5 peptide panel. One peptide in the N-terminal fragment of M5 protein and one peptide in the C-terminal fragment significantly stimulated the lymphocytes. However, these T cell epitopes were found to be different amino acid sequences from those peptides recognized by antibodies. Similar results were also found for the myosin crossreactive T cell epitopes of M5 protein (Antone and Cunningham, in preparation).

The studies described above reflect the complexity of the immune system and suggest that host response to particular epitopes may determine the role each will play in the production of disease. Subtle differences in individual host responsiveness to certain epitopes may even determine the type of manifestations present in the disease. In the case of cytolytic antibodies such as mAb 36.2.2, only certain hosts may have the potential to make these antibodies. Acute rheumatic fever sera do block the reaction of mAb 36.2.2 with specific M5 peptides (Antone and Cunningham, unpublished data), and normal sera do not contain such antibodies that block the binding of mAb 36.2.2. In the future, crossreactive T cell clones will need to be developed in order to study the role of T cells in rheumatic heart disease.

Acute Rheumatic Fever Linked to Systemic Lupus Erythematosus and Sjogren's Syndrome by Anti-Myosin Idiotype

Because anti-myosin antibodies were associated with crossreactivity between the group A streptococcus and heart, an anti-idiotypic antiserum was produced to study anti-myosin antibodies in streptococcal diseases and sequelae. The anti-idiotypic antiserum was used to determine if acute rheumatic fever might be connected to other autoimmune diseases through an anti-myosin antibody idiotype. An anti-idiotypic serum was made in rabbits by immunizing them with affinity purified anti-myosin antibodies from acute rheumatic fever. The serum was rendered idiotype specific by absorption with normal immunoglobulin depleted of anti-myosin antibody (McCormack *et al.*, submitted). Specificity was determined by the reactivity of absorbed rabbit anti-idiotypic serum with affinity purified immunoglobulin from acute rheumatic fever serum. No reactivity was observed with affinity purified immunoglobulins from normal sera. The anti-idiotypic reagent reacted with affinity purified anti-myosin antibodies from positive sera (McCormack *et al.*, submitted) and effectively inhibited the binding of mouse or human crossreactive mAbs to whole streptococci or myosin in the ELISA. These data suggested that the idiotype was present on both mouse and human antibodies.

When anti-idiotypic serum was reacted with antibodies from other autoimmune diseases, anti-myosin antibody idiotype was found to be elevated in systemic lupus erythematosus, Sjogren's syndrome, and post-streptococcal acute glomerulonephritis as well as in acute rheumatic fever (McCormack *et al.*, submitted). Sera from chronic autoimmune myocarditis, heart failure, rheumatoid arthritis, IgA nephropathy, uncomplicated streptococcal infection, and Chagas disease all contained normal levels of the idiotype.

The postinfectious nature of acute rheumatic fever and acute glomerulonephritis is well-documented, while the role of infectious agents in systemic lupus erythematosus and Sjogren's syndrome is not as well-defined. Further studies will investigate the role of anti-myosin antibody idiotype in these diseases and attempt to understand their relationship.

Epitopes Shared Between Viruses and Streptococcal M Proteins

Both rheumatic carditis and autoimmune myocarditis are associated with the presence of anti-myosin antibodies (Cunningham *et al.*, 1988 and Wolfgram *et al.*, 1985; Neu *et al.*, 1987 a,b). Clinically the two diseases are distinct (Stollerman, 1989), but both are the result of previous infections (Rose, 1988). Since viruses are a recognized etiologic agent of myocarditis, we tested the hypothesis that viruses might share epitopes with group A streptococci. Viruses have also been reported to share epitopes with host proteins (Fujinami *et al.*, 1989) and have been suggested as an environmental trigger of autoimmune diseases.

Anti-streptococcal mAbs crossreactive with M protein and myosin were tested for neutralization of coxsackieviruses and poliovirus type 1. Table IV summarizes the data. MAbs 36.2.2, 54.2.8 and 49.8.9 neutralized either coxsackievirus group B or poliovirus type 1. Several other IgM mAbs tested, including 654.1.1, were found to be non-neutralizing (Cunningham *et al.*, 1992). In Western blots of coxsackievirus B3m, mAb 36.2.2 recognized the viral capsid protein, VP3, while 54.2.8 reacted with VP1. MAb 49.8.9 neutralized most of the coxsackie viruses tested and reacted with all three of the viral capsid proteins in Western blot (Cunningham *et al.*, 1992).

Table IV

Reactivity of anti-streptococcal monoclonal antibodies with viruses in a plaque reduction assay

Mab	Coxsackievirus			Poliovirus
	B4	B3m	B3as	
36.2.2	-	+/-	-	+
54.2.8	-	+/-	+/-	+
49.8.9	+	+	-	-
654.1.1	-	-	-	-

- = values below 25% plaque reduction;
+/- = values between 25-50% plaque reduction;
\+ = values between 50-100% plaque reduction.

Table V

Overlapping sequence homology between coxsackie virus VP1, streptococcal M protein and human cardiac myosin

	Human Cardiac Myosin	Streptococcal M Protein
Coxsackie B3 VP1 (88-106)	1552-1566 (40% identity in a 15 amino acid overlap)	370-383 (40% identity in a 14 amino acid overlap)
Poliovirus VP1 (218-243)	1651-1678 (28.6% identity in a 28 amino acid overlap)	0

Significant sequence homology was found to exist between streptococcal M6 protein, coxsackievirus B3m VP1 protein, and human cardiac myosin as shown in Table V. No significant sequence homology was found between poliovirus VP1 and M6 protein (Cunningham *et al.*, 1992). The shared sequence in human cardiac myosin was located in the α−helical rod in the light meromyosin tail fragment (Figure 3). VP1 of coxsackievirus was 40% homologous (Table V) with M protein and myosin in the α−helical segment of the capsid protein (Cunningham *et al.*, 1992). The overlapping region in M 6 protein (Table V) was recently shown to share 31% identity with human cardiac tropomyosin (Fenderson *et al.*, 1989). The site of homology in M protein is located in the carboxy terminal C- and D- repeat regions of the molecule (Fischetti *et al.*, 1988).

Most importantly, anti-streptococcal crossreactive mAbs neutralized certain viruses and may be involved in protection of the host against infection as well as in destruction of tissues. The host may take advantage of shared α-helical coiled-coil structures found in many bacterial and viral proteins (Cohen and Phillips, 1981) and respond to a common determinant which may protect the host against a variety of pathogens. It is possible that such promiscuous immune responses are primordial in origin.

Overview

The data presented in this chapter demonstrate that heart-reactive antibodies which recognize both group A streptococci and heart tissue have interesting characteristics. The crossreactivity was broad-based and encompassed a number of related antigens primarily containing α−helical coiled-coil structures including myosin, tropomyosin, keratin, vimentin and laminin. The crossreactive epitopes in viruses have not yet been defined but may be related to the α−helical coiled-coil regions in capsid proteins. The crossreactive antibodies which are lupus-like and recognize DNA are similar to anti-nuclear antibodies reported previously (Lafer *et al.*, 1981; Shoenfeld and Schwartz, 1984).

Study of anti-myosin antibodies has led to the discovery that acute rheumatic fever may be related to systemic lupus erythematosus and Sjogren's syndrome as well as to acute poststreptococcal nephritis. The challenge will be to determine how these diseases are related but different. One common thread might be their hyperresponsiveness to common α-helical coiled-coil structures of viruses and bacteria.

The role of these crossreactive antibodies in acute rheumatic fever and heart disease is still unclear. However, their reactivity with valve components has led to the hypothesis that they may damage the outer endothelium of the valve and react with underlying constituents (Gulizia *et al.*, 1991). The reaction of the valvular interstitial cell with these antibodies may suggest that these cells play a role in the overall inflammation and collagen deposition in the valve. The crossreactive antibodies may become cytotoxic for heart cells if they recognize a cell surface component and may play a role in development of rheumatic carditis.

Our findings present some important clues for understanding the potential role of immunological crossreactivity in acute rheumatic fever and heart disease as well as in other autoimmune and rheumatic diseases.

Acknowledgements

Gratitude is expressed to all of my students and fellows who have made this work possible; to Carol Crossley for excellent technical assistance; to Mary Patterson and Rose Haynes for excellent secretarial assistance; to Drs. Joseph Ferretti, Vincent Fischetti, the late Edwin Beachey, James Dale, William Clark, Bruce McManus, and Charles Gauntt for providing helpful reagents; to Drs. Noel Rose, Ahvie Hersowitz, Steven Reed, Susan Jackson, L.T. Chun, D.V. Reddy and Elia Ayoub for kindly providing human sera; to The Presbyterian Health Foundation for equipment, and to my three children, Catherine, Nicole and Luke, who waited patiently while this work was done. This work was supported by grants from the American Heart Association #84-1337 to M.W.C. and #OK-87-F-5 to V.A.D. and from the National Heart, Lung and Blood Institute #HL35280 and #HL01913. M.W.C. is the recipient of an NIH Research Career Development Award.

References

Antone, S.M. and Cunningham, M.W., Cytotoxic anti-streptococcal monoclonal antibody recognizes laminin, In preparation.

Antone, S.M. and Cunningham, M.W., Myosin and laminin crossreactive T cell epitopes of streptococcal M5 protein, In preparation.

Ayakawa, G.Y., *et al.*, 1988, Heart cross-reactive antigens of mutans streptococci share epitopes with group A streptococci and myosin, *J. Immunol.* 140:253-257.

Ayoub, E.A., *et al.*, 1986, Association of class II human histocompatability leukocyte antigens with rheumatic fever, *J. Clin. Invest.* 77:2019-2026.

Barnett, L.A. and Cunningham, M.W., 1990, A new heart crossreactive antigen in *Streptococcus pyogenes* is not M protein, *J. Infect. Dis.* 162:875-882.

Barnett, L.A. and Cunningham, M.W., Evidence for actin-like and myosin-like proteins in *Streptcoccus pyogenes*, *Infect. & Immun.*, In press.

Bessen, D., Jones, K.E., and Fischetti, V.A., 1989, Evidence for two distinct classes of M protein and their relationship to rheumatic fever, *J. Exp. Med.*, 169:269-283.

Bisno, A.L., 1979a, Classification of streptococci, In: Principles and Practices of Infectious Diseases, Mandell, G.L., Douglas, Jr., R.G., Bennett, J.E. (eds.), Wiley & Sons, Inc. New York, pp. 1559-1573.

Bisno, A.L., 1979b, Rheumatic fever, In: Principles and Practices of Infectious Diseases, Mandell, G.L., Douglas, Jr., R.G., Bennett, J.E. (eds.), Wiley & Sons, Inc. New York, pp.1574-1582.

Bisno, A.L., 1979c, Glomerulonephritis, In: Principles and Practices of Infectious Diseases, Mandell, G.L., Douglas, Jr., R.G., Bennett, J.E. (eds.), Wiley & Sons, Inc. New York, pp.1583-1589.

Bronze, M.S., Beachey, E.H., and Dale, J.B., 1988, Protective and heart crossreactive epitopes located within the NH_2 terminus of type19 streptococcal M protein, *J. Exp. Med.* 167:1849-1859.

Chalovich, J.M. and Fischetti, V.A., 1986, Crosslinking of actin filaments and inhibition of actomyosin subfragment-1 ATPase activity by streptococcal M6 protein. *Arch. Biochem. Biophys.* 245:37-43.

Clark, W.A., *et al.*, 1984, Classification and characterization of cardiac isomyosins, *Eur. Heart J.* 5:69-75.

Clarke, M. and Spudich, J.A., 1974, Biochemical and structural studies of actomyosin-like proteins from non-muscle cells. Isolation and characterization of myosin from amoebae of Dictyostelium discoideum, *J. Mol. Biol.* 86:209-222.

Cohen, C. and Phillips, G.N., 1981, Spikes and fimbriae: α−helical proteins form surface projections on microorganisms, *Proc. Natl. Acad. Sci. USA*, 78:5303-5304.

Cunningham, M.W., 1992, Cytotoxic and viral neutralizing antibodies crossreact with streptococcal M protein, coxsackieviruses, and human cardiac myosin, *Proc. Natl. Acad. Sci. USA*, 89: 1320-1324.

Cunningham, M.W., Graves, D.C. and Krisher, K., 1984, Murine monoclonal antibodies reactive with human heart and group A streptococcal membrane antigens, *Infect. Immun.* 46:34-41.

Cunningham, M.W. and Russell, S.M., 1983, Study of heart reactive antibody in antisera and hybridoma culture fluids against group A streptococci, *Infec. Immun.* 42:531-538.

Cunningham, M.W. and Swerlick, R.A., 1986, Polyspecificity or antistreptococcal murine monoclonal antibodies and their implications in autoimmunity, *J. Exp. Med.* 164:998-1012.

Cunningham, M.W., *et al.*, 1986, A study of anti-group A streptococcal monoclonal antibodies crossreactive with myosin, *J. Immunol.* 136:293-298.

Cunningham, M.W., *et al.*, 1988, Human monoclonal antibodies reactive with antigens of the group A streptococcus and human heart, *J. Immunol.* 141:2760-2766.

Cunningham, M.W., *et al.*, 1989, Human and murine antibodies crossreactive with streptococcal M protein and myosin recognize the sequence gln-lys-ser-lys-gln in M protein, *J. Immunol.* 144:2677-2683.

Dale, J.B. and Beachey, E.H., 1985a, Epitopes of streptococcal M proteins shared with cardiac myosin, *J. Exp. Med.* 162:583-591.

Dale, J.B. and Beachey, E.H., 1985b, Multiple heart crossreactive epitopes of streptococcal M proteins, *J. Exp. Med.* 161:113-122.

Dale, J.B. and Beachey, E.H., 1986, Sequence of myosin crossreactive epitopes of streptococcal M protein, *J. Exp. Med.* 164:1785-1790.

Dell, V.A., *et al.*, (1991) Autoimmune determinants of rheumatic carditis: Localization of epitopes in human cardiac myosin, *Eur. Heart J.* 12: (Supplement D) 158-162.

Eisenberg, B.R., 1987, Relationship of membrane systems in muscle to isomyosin content, *Conf. J. Physiol. Pharm.* 65:598-605.

Fenderson, P.G., Fischetti, V.A., and Cunningham, M.W., 1989, Tropomyosin shares immunologic epitopes with group A streptococcal M proteins, *J. Immunol.* 142:2475-2481.

Ferretti, J.J., Shea, C., and Humphrey, M.W., 1980, Cross-reactivity of Streptococcus mutans antigens and human heart tissue, *Infect. Immun.* 30:69-73.

Fischetti, V.A., Hodges, W.M., and Hruby, D.E., 1989, Protection against streptococcal pharyngeal colonization with a vaccinia: M protein recombinant, *Science*, 244:1487-1490.

Fischetti, V.A. and Manjula, B.N., 1982, Biologic and immunologic implications of the structural relationship between streptococcal M protein and mammalian tropomyosin, In: Seminars in Infections Disease, Bacterial Vaccines, Vol. 4, Robbins, J.B., Hill, J.C., and Sadoff J.C. (eds) ,Theime Stratton Inc., New York pp. 411-418.

Fischetti, V.A., *et al.*, 1988, Conformational characteristics of the complete sequence of group A streptococcal M6 proteins, *Proteins Struct. Funct. Genet.* 3:60-69.

Froude, J., *et al.*, 1989, Cross-reactivity between streptococcus and human tissue: A model of molecular mimicry and autoimmunity, Current Topics in *Microbiol. Immunol.* 145:5-26.

Fujinami, R.S., 1988, Viruses and molecular mimicry, In: Molecular Mimicry in Health and Disease eds. Lernmark, A., Dynberg, T., Terenius, L. and Horbelt, B. (eds), Elsevier Science Publishers, Amsterdam, The Netherlands, pp. 237-244.

Gulizia, J.M., Cunningham, M.W., and McManus, B.M., 1991, Immunoreactivity of anti-streptococcal monoclonal antibodies to human heart valves: Evidence for multiple crossreactive epitopes, *Amer. J. Pathol.,* 138: 285-301.

Hollingshead, S.K., Fischetti, V.A., and Scott, J.R., 1987, A highly conserved region present in transcripts endoding heterologous M proteins of group A streptococci, *Infect. Immun.* 55:3237-3239.

Kaplan, M.H., 1963, Immunologic relation of streptococcal and tissue antigens. I. Properties of an antigen in certain strains of group A streptococci exhibiting immunologic crossreaction with human heart tissue, *J. Immunol.* 90:595-606.

Kaplan, M.H., *et al.*, 1964, Presence of bound immunoglobulins and complement in the myocardium in acute rheumatic fever, *N. Engl. J. Med.* 271:637-645.

Khanna, A.K., *et al.*, 1989, The presence of a non-HLA B-cell antigen in rheumatic fever patients and their families as defined by monoclonal antibody, *J. Clin. Invest.* 83:1710-1736.

Kraus, W., Seyer, M., and Beachey, E.H., 1989, Vimentin-cross-reactive epitope of type 12 streptococcal M protein, *Infect. Immun.* 57:2457-2461.

Kraus, W., *et al.*, 1989, Autoimmune sequence of streptococcal M protein shared with the intermediate filament protein, vimentin, *J. Exp. Med.* 169:481-492.

Krisher, K. and Cunningham, M.W., 1985, Myosin: A link between streptococci and heart, *Science* 227:413-415.

Lafer, E.M., *et al.*, 1981, Polyspecific monoclonal lupus autoantibodies reactive with both polynucleotides and phosphalipids, *J. Exp. Med.* 153:897-909.

Lancefield, R.C., 1962, Current knowledge of type-specific M antigens of group A streptococci, *J. Immunol.* 89:307-313.

Manjula, B.N., Trus, B.L., and Fischetti, V.A., 1985, Presence of two distinct regions in the coiled-coil structure of streptococcal pep M5 protein: relationship to mammalian coiled-coil proteins and implications to its biological properties, *Proc. Natl. Acad. Sci. USA* 82:1064-1068.

McCormack, J.M., *et al.*, Systemic lupus erythematosus and Sjogren's syndrome linked to streptococcal sequelae by anti-myosin idiotype, Submitted.

Mische, S.M., Manjula, B.N., and Fischetti, V.A., 1987, Relation of streptococcal M protein with human and rabbit tropomyosin: The complete amino acid sequence of human cardiac alpha tropomyosin, a highly conserved contractile protein, *Biochem. Biophys. Res. Comm.* 142:813-818.

Moore, P.B., Huxley, H.E., and deRosier, D.J., 1970, Three-dimensional reconstruction of F-actin, thin filaments and decorated thin filaments, *J. Mol. Biol.* 50: 279-295.

Neu, N., *et al*, 1987a, Cardiac myosin induces myocarditis in genetically predisposed mice, *J. Immunol.* 139:3630-3636.

Neu, N., *et al.*, 1987b, Autoantibodies specific for the cardiac myosin isoform are found in mice susceptible to Coxsackievirus B3-induced myocarditis, *J. Immunol.* 138:2488-2492.

Patarroyo, M.E., *et al.*, 1979, Association of B cell alloantigen with susceptibility to rheumatic fever, *Nature*, 278:173-174.

Read, S.E., *et al*, 1974, Cellular reactivity studies to streptococcal antigens. I. Migration inhibition studies in patients with streptococcal infections and rheumatic fever, *J. Clin. Invest.* 54:439-450.

Reid, H.M.F., *et al.*, 1980, Lymphocyte response to streptococcal antigens in rheumatic fever patients in Trinidad. In: Streptococcal diseases and the immune response, Read, S.E. and Zabriskie, J.B. (eds.), Academic Press, New York, pp.681-693.

Rose, N.R., Herskowitz, A., Neumann, D.A., and Neu, N., 1988, Coxsackie B3 infection and autoimmune myocarditis, In: Molecular Mimicry in Healtheand Disease, Lernmark, A., Dyrberg, T.,Terenius, L., and Hokfelt, B., (eds.) Elsevier Science Publishers B.V.(Biomedical Division), Amsterdam, The Netherlands. pp.273-284.

Rose, N.R., Wolfgram,L.J., Herskowitz,A., and Beisel,K.W., 1986, Postinfectious autoimmunity: Two distinct phases of Coxsackievirus B3-induced myocarditis, In: Autoimmunity: Experimental and Clinical Aspects, Vol. 475, Schwartz, R.S., and Rose, N.R. (eds.), New York Academy of Sciences, pp. 146-156.

Sargent, S.J., *et al.*, 1987, Sequence of protective epitopes of streptococcal M protein shared with cardiac sarcolemmal membranes, *J. Immunol.* 139:1285-1290.

Shoenfeld, Y. and Schwartz, R.S., 1984, Immunologic and genetic factors in autoimmune diseases, *N. Eng. J. Med.* 311:1019-.1029

Shulman,S.T.,and Ayoub, E.M., 1974, Qualitative and quantitative aspects of the human antibody response to streptococcal group A carbohydrate, *J. Clin. Invest.* 54:990-996.

Stollerman, G.H., 1988, Rheumatic and heritable connective tissue diseases of the cardiovascular system, In: Heart disease: A textbook of cardiovascular medicine, Vol. 2, Braunwald, E. (ed.), W.B. Saunders Co., Philadelphia, PA, pp. 1710-1733.

van de Rijn, I., Bleiweis, A.S., and Zabriskie, J.B., 1076, Antigens in Streptococcus mutans cross-reactive with human heart muscle, *J. Dent. Res.* 55C:59-64.

van de Rijn, I., Zabriskie, J.B., and McCarty, M., 1977, Group A streptococcal antigens crossreactive with myocardium. Purification of heart reactive antibody and isolation and characterization of the streptococcal antigen. *J. Exp. Med.* 146:579-599.

Wolfgram, L.J.., Beisel, K.W., and Rose, N.R., 1985, Heart-specific autoantibodies following murine Coxsackievirus B3 myocarditis, *J. Exp. Med.* 161:1112-1121.

Zabriskie, J.B. and Freimer, E.H., 1966, An immunological relationship between the group A streptococcus and mammalian muscle, *J. Exp. Med.* 124:661-678.

Zabriskie, J.B., Hsu, K.C., and Seegal, B.C., 1970, Heart reactive antibody associated with rheumatic fever: Characterization and diagnostic significance, *Clin. Exp. Immunol.* 7:147-159.

Attenuated *Salmonella sp.* as Live Vaccines and as Presenters of Heterologous Antigens

Bruce A.D. Stocker

Introduction

The first bacterial vaccines were Pasteur's attenuated strains of *Pasteurella multocida* and *Bacillus anthracis*, but until recently there has been relatively little effort to develop or use such strains for protection of humans, with the conspicuous exception of strain BCG, for prevention of tuberculosis.

Recent interest in the development and trial of candidate live-vaccine strains of *Salmonella typhi* results, I think, from several factors: (i) Dissatisfaction with the killed whole-cell vaccines, given by injection, because controlled field trials have showed them to be only moderately effective (between 53 and 79% efficacy in most trials) and because of the very high incidence of side-effects (local discomfort, fever) unpleasant enough to prevent their routine use in civilian populations; (ii) Realization, from study of model *Salmonella sp.* infections in mice, that killed vaccines, though they may cause good antibody production, confer much less protection than does a non-fatal infection or the administration of a safely attenuated strain and that this reflects the ability of the live bacteria to cause cellular immune responses not achievable by use of killed vaccine; (iii) Increased understanding of bacterial physiology and pathogenesis, allowing a better choice of characters for attenuation, together with increased ability to construct strains with desired characters, by the techniques of bacterial genetics and, more recently, by recombinant DNA methods.

Criteria for a Live Typhoid Vaccine

The characters needed for a safe and effective live-vaccine strain of *S. typhi* appear to be: (i) Irreversible genetic lesion(s) such that the bacteria cannot multiply in human tissues for more than a very few generations (and so cannot cause a systemic infection); (ii) Retention of wild-type antigens and virulence factors, to increase the probability of protective efficacy; (iii) Ability of the live-vaccine bacteria to survive for at least several days in the tissues of the vaccinated subject and that this survival should be intracellular (an attenuated strain of *S. enteritidis* otherwise effective as live vaccine gave no protection when administered in an implanted chamber, allowing passage of macromolecules but preventing contact of bacteria and host cells Akiyama *et al*., 1964); (iv) No or only trivial side-effects or reactions, local or general.

The requirements for a satisfactory live vaccine for use in animals are much the same, but perhaps less stringent. Strains of *Salmonella sp*. attenuated by different methods and tested or in use to prevent disease in animals include those with not fully characterized defects in lipopolysaccharide (LPS) structure (Smith, 1965); with uncharacterized mutations preventing growth at 37°C (Fahey and Cooper, 1970); with various auxotrophic characters (Hoiseth and Stocker, 1981; Linde, 1983; Linde *et al*., 1990); and those resistant to antibiotics by mutations found to cause also reduced virulence (Linde, 1983; Linde *et al*., 1990) or antibiotic dependence, discussed below.

Streptomycin-dependent and galE Strains as Live Vaccines

Several rationally designed candidate live-vaccine strains of *S. typhi* have been developed and tested in man. Perhaps the earliest was the streptomycin-dependent mutant tested in the early seventies. Given to volunteers in doses up to 10^{11} CFU, it proved harmless and protective, as tested by challenge, in two trials (Levine *et al*., 1976). In two other trials the vaccine, given as freshly rehydrated freeze-dried material instead of as broth culture, gave no protection, and the strain was not further investigated. An argument against use of this sort of strain is that the molecular nature of its mutation is not known and that reversion of streptomycin-dependent mutants to independence, though generally not to virulence, has been observed in other bacterial genera.

The best-known *S. typhi* live vaccine is strain Ty21a, developed by the late Dr. Rene Germanier, of the Swiss Serum Institute, as a galE mutant, lacking the enzyme which interconverts UDPgalactose and UDPglucose. Such mutants cannot make UDPgalactose unless provided an exogenous supplye of galactose (and cannot utilize galactose as a source of energy). Failure to make UDPgalactose prevents formation of the galactose units of the LPS core and of each O repeat unit resulting in cells with a "rough" phenotype, and incomplete core LPS of type Rc. galE mutants which retain function of the other two genes of the gal operon, galK and galT, assimilate galactose and convert some of it, via galactose phosphate, to UDPgalactose, allowing synthesis of normal LPS, and a switch to "smooth" phenotype. However such galE bacteria lyse within an hour or so of exposure to galactose, perhaps as a result of accumulating galactose phosphate and UDPgalactose. In testing different sorts of rough mutants of *S. typhimurium* in mice Germanier found that a galE mutant was attenuated to about the same extent as mutants with other LPS core defects, but differed by the substantial protection it conferred against later challenge (Germanier, 1970; Germanier and Furer, 1971). galE mutants of *S. typhimurium* have proven effective as live vaccines in mice; but some galE mutants in another serotype retained virulence (Nnalue and Stocker, 1986) and an inoculum which was harmless in a normal animal caused death in mice treated with cyclophosphamide (Morris *et al.*, 1976).

A galE mutant of *S. typhi* designated Ty21a, obtained after exposure of a culture to nitrosoguanidine, appeared appropriately attenuated and immunogenic so far as could be tested in small animals and, at least in its final form, did not revert to gal^{+} at detectable frequency (Germanier and Furer, 1975). In tests at the Center for Vaccine Development at the University of Maryland School of Medicine it was found harmless when given by mouth to volunteers. Later challenge showed that it resulted in substantial protection (Gilman *et al.*, 1977). It was sensationally effective, giving a 97% reduction in incidence in school children observed over three years, in a first field trial, in Egypt (Wahdan *et al.*, 1982). In later extensive field trials, in Chile, conducted by Dr. M. Levine and his colleagues at Baltimore and Chile, strain Ty21a, given in various forms and dosage schedules, has proven harmless in more than 600,000 children, and has given protection, though of variable extent, usually between 40 and 70%, but never as good as that seen in the trial in Egypt (Levine *et al.*, 1990). Use of this strain is now licensed in many countries, including the U.S. It is safe, moderately effective, and free of unpleasant side-effects.

Strain Ty21a has several undesirable properties, presumably resulting from mutations caused by the mutagen exposure involved in its construction. It lacks the Vi antigen, known to be a protective factor. It has new nutritional requirements and grows much more slowly than its wild-type parent even on rich media; and it is difficult to freeze-dry in a satisfactory form. Dr. David Hone and his colleagues, in Australia, hoping to make a better galE live-vaccine strain, constructed, by recombinant DNA methods, a strain with an extensive deletion in the galE gene. Then, as an additional precaution, they isolated a Vi-negative mutant of it. Tests *in vitro* and in mice indicated the desired properties. However, despite these precautions, two of the four volunteers (including one of the scientists concerned in the project) who ingested $7x10^{8}$ CFU fell sick 5 or 6 days later with fever and positive blood cultures (Hone et al., 1988). The re-isolated strain was indistinguishable from the strain as administered. It was concluded that a complete block at galE, together with loss of Vi antigen, does not substantially attenuate *S. typhi*, as judged by the dose sufficient to cause disease in man.

This and other evidence indicate that the non-virulence of Ty21a results, entirely or in large part, from one or more mutations, of unknown nature, but not from its galE defect. A Vi^{+} derivative of Ty21a has been made by integration of the cloned wild-type gene into the chromosome (Cryz *et al.*, 1989). This may improve its immunogenic potential. Curing of the galE defect of Ty21a would perhaps not compromise its non-virulence, might improve its ability to cause O-specific responses and would eliminate the need for precautions in manufacture to ensure a good yield of phenotypically smooth bacteria. However, the great expense of volunteer trials conducted in isolation facilities would probably prevent test of this surmise, as of many other trials which might hasten the advent of a fully effective vaccine for prevention of the several millions of cases of typhoid fever occuring each year, world-wide.

Aromatic-dependent *Salmonella* as Live Vaccine

In the early fifties, an investigation (Bacon *et al.*, 1951) of the effect of mutation to auxotrophy on the ability of *S. typhi*, injected i.p. without adjuvant, to cause fatal infections in mice showed that a mutant requiring p-aminobenzoic acid (pAB) was less virulent than its wild-type parent. This reduced virulence was explicable by the absence of pAB in mammalian tissues, the only role of this compound being as precursor of folic acid, which vertebrates acquire, preformed, in their diet. By contrast, bacteria and plants must make all essential aromatic metabolites (except any supplied from the exterior) from chorismic acid, the final product of the common aromatic biosynthesis pathway. Some ten years ago it occurred to me that blocking this pathway would make *Salmonella* dependent on an external supply of aromatic metabolites, including pAB, and that such bacteria would be able to multiply for at most only a very few generations in host tissues, and consequently would be unable to cause systemic infections. (Another product of chorismate is 2,3-dihydroxybenzoic acid, the precursor of enterobactin, used by *Salmonella*, etc., for capturing ferric iron and, like pAB, absent in host tissues; however, inability to make enterobactin does not significantly attenuate *S. typhimurium*, as tested by i.p. injection in BALB/c mice, S.K. Hoiseth and Stocker, unpublished). The discovery of transposons and availability of *S. typhimurium* LT2 lines with transposon Tn10, which confers resistance to tetracycline, inserted in gene aroA (whose product catalyzes a step in the aromatic biosynthesis pathway) allowed a test of the attenuating effect of a block in this path. Susan Hoiseth and I found that virulent strains of *S. typhimurium* and *S. dublin* made aroA::Tn10 by transduction with selection for tetracycline-resistance were profoundly attenuated, e.g., LD_{50}, i.p., for BALB/c mice greater than 10^6 CFU, compared to less than 20 for the parent strain. Such aroA::Tn10 strains can revert to aromatic independence, and so to virulence, by the rare event of "clean excision" of the transposon; this possibility can be eliminated by selection of tetracycline-sensitive mutants arising by transposon-generated deletions or inversions extending from the transposon into or through aroA coding sequence, making reversion impossible.

Stable aroA strains thus obtained were very effective as live vaccines in mice, a single i.p. inoculum or a single feeding protecting mice against parenteral or oral challenge with a virulent strain, given in numbers corresponding to tens of thousands of i.p. LD_{50} doses or to hundreds of oral LD_{50} doses (Hoiseth and Stocker, 1981). Strains similarly derived from several *Salmonella* serotypes, from my laboratory or constructed elsewhere, have given generally similar results when tested as live vaccines in mice, calves and sheep (Smith *et al.*, 1984; Robertsson *et al.*, 1983; Mukkur *et al.*, 1987).

Auxotrophic *S. typhi* as Live Vaccine

Mary Frances Edwards and I used this information for construction of candidate live-vaccine strains of *S. typhi* (Edwards and Stocker, 1988). A previously characterized extensive deletion at locus aroA was introduced into the chromosome of each of two wild-type strains by a procedure involving use of a Tn10 insertion affecting gene serC, in the same operon as aroA and promoter-proximal to it (Hoiseth and Stocker, 1985). Each ΔaroA transductant was then given point mutation hisG46 by two steps of transduction to provide an extra label to distinguish the live-vaccine strain from wild strains. It then seemed desirable that a strain to be given to volunteers should have a second, independent, attenuating character, due to mutation at a point on the chromosome well-separated from aroA. Another investigation (McFarland and Stocker, 1987) had shown that the attenuating effect of mutation to purine requirement, originally reported by Bacon *et al.* (1951), differed between nutritional classes; only blocks at purA or purB, each causing requirement for adenine or adenosine, resulted in complete loss of virulence in *S. dublin* and *S. typhimurium*, as tested in mice. We were able to introduce a proven deletion at purA into *S. typhi* CDC10-80 already made aroA hisG46. The final product, with two deletion mutations, each expected to result in complete inability to cause typhoid fever, that is strain 541Ty, and its Vi-negative mutant 543Ty, were tested in volunteers by Dr. Mike Levine and his colleagues at the Center for Vaccine Development, University of Maryland School of Medicine (Levine *et al.*, 1987). Each strain proved safe and caused no side-effects, even in volunteers taking 2×10^{10} CFU, preceded by bicarbonate, to prevent killing by gastric acid. All the volunteers given the larger dose gave evidence of cellular immune responses, which might have been taken as predicting protection. However, challenge with a virulent strain is now considered inpermissible. To our disappointment, very few of the volunteers showed any humoral antibody response. Two later investigations (Sigwart *et al.*, 1989; O'Callaghan *et al.*, 1988) on the effect of purA mutation in *S. typhimurium* or *S. dublin* showed that this character much reduced the survival of aroA bacteria in mouse liver and spleen, and correspondingly reduced their immunogenic and protective efficacy. To avoid this effect and yet retain the advantage of having two independent attenuating mutations, it was proposed (Stocker, 1988) to construct new candidate strains, with complete blocks at two different points in the aromatic biosynthesis pathway, by mutations at well-separated loci. A strain with "engineered" deletions in two aro genes, constructed by Dr.

David Hone at the Center for Vaccine Development, has recently been tested in volunteers. The results of this trial are not yet published, but I have permission from Drs. Hone and Levine to mention that they consider them encouraging.

Other candidate strains of the same sort have been or are being constructed, and I think that such strains are likely to prove satisfactory and safe when tested, first in volunteers, later in field trials. Strains of *S. typhi* attenuated by other mechanisms, such as the combination of stable cya and crp mutations (Curtiss and Kelly, 1987), may also prove effective, since they have given satisfactory protection in animal models using *S. typhimurium*, instead of *S. typhi*. Furthermore, a product vaccine comprising purified Vi polysaccharide has proven, in first field trials, about as effective as Ty21a (Robbins and Schneerson, 1990).

Salmonella Live-Vaccine Strains as Presenters of Foreign Antigens

One of the first antigen-determining plasmids introduced into a live-vaccine strain was the 140 MD non-conjugative plasmid present in all Form-1 (smooth) strains of *Shigella sonnei*; this carries all the genes needed for synthesis of the O side-chain of the LPS of this species. Dr. S. Formal and his colleagues at the Walter Reed Army Institute of Research were able to transfer a recombinant form of this plasmid by conjugation into *S. typhi* Ty21a, described above. The O antigen of *Sh. sonnei* was expressed at the bacterial surface of the transconjugant but as a macromolecule not covalently joined to the LPS core. This strain was fed to volunteers in several trials. The later administration of wild-type *Sh. sonnei* showed that vaccine had in some but not in all trials conferred partial protection (Herrington et al., 1990). Derivatives of Ty21a expressing O antigens of *Sh. flexneri* or of *Sh. dysenteriae* 1 have also been made by transfer of constructed plasmids which include the relevant genes specifying enzymes needed for biosynthesis of these polysaccharides; no reports of their trial in volunteers have as yet appeared. The O antigens of both *Vibrio cholerae* and of *S. typhi* were expressed by strain Ty21a with its LPS core structure appropriately modified and given constructed plasmids containing relevant genes from *Vibrio cholerae* (Tacket et al., 1990). All of ten volunteers fed this strain made IgA reactive with *Salmonella* O antigen and seven of the ten also made antibody reactive with *V. cholerae* O antigen.

The use of Ty21a as presenter of foreign antigens has the advantage that the resulting strain can be tested in volunteers, for possible utility as immunizing agent for use in humans, but has the disadvantages resulting from its galE character and some of its pecularities mentioned above. Many investigators have used, instead, aromatic-dependent strains or galE strains of *Salmonella sp*. pathogenic for mice, in particular *S. typhimurium* and *S. dublin*, as hosts for constructed plasmids specifying production of heterologous proteins of interest as model antigens or because identified as protective antigens. Brown *et al.* (1987) used a aroA live-vaccine strain of *S. typhimurium*, SL3261, as host of a plasmid which included a cloned lacZ gene, resulting in constitutive production of β-galactosidase (because *Salmonella*, which do not make this enzyme, have neither the lacZ operon nor its repressor gene, lacI). Mice given this construct as live vaccine by the i.v. route made an anti-LacZ antibody response and also developed delayed-type sensitivity to the purified protein. An aroA strain of *S. dublin*, SL1438, used as carrier of a plasmid causing production of the B subunit of the heat-labile enterotoxin of *E. coli*, administered p.o., 2 doses, each 10^{10} CFU, to BALB/c mice led to increasing titers of serum and gut antibody, both to *Salmonella* LPS and to the B subunit (Clements et al., 1986). Gordon Dougan and his colleagues at Wellcome Research Laboratories have used SL3261 and other aromatic-dependent live vaccine strains as carriers of plasmids specifying a variety of proteins, including *E. coli* heat-labile enterotoxin subunit B, influenza virus nucleoprotein, a fragment of tetanus toxin and a Leishmania antigen and have obtained immune responses, not only antibody production but in some systems also cytotoxic activity and other evidence of T cell responses (Maskell *et al.*, 1987; Tite *et al.*, 1988; Fairweather *et al.*, 1990). In a system in which an adequate antibody response was expected to be sufficient for protection a plasmid including the gene for the mature form of the M protein of *Streptococcus pyogenes* type 5 was placed in an aromatic-dependent *S. typhimurium* strain, SL3261, and the resulting strain used to immunize mice by feeding two doses, each 10^9 CFU. The mice made an antibody response to the M protein and were protected against i.p. or intranasal challenge with type 5 *Streptococcus pyogenes*, sufficient to kill the non-vaccinated control mice (Poirier *et al.*, 1988).

Immune Response to Epitopes Inserted in *Salmonella* Flagellin

Another way of using *Salmonella* to obtain an immune response is the insertion into a cloned flagellin gene of a synthetic oligonucleotide specifying a short amino acid sequence, known or suspected to be an epitope of some heterologous protein antigen -- a procedure analogous to the use of gene lamB of *E. coli* as target for insertion of epitope-specifying sequences by Maurice Hofnung and his colleagues (Charbit *et al.*, 1986, 1988). This system, developed by Dr. Salete Newton (Newton *et al.*, 1989, 1990) and at Praxis Biologics, Inc. (Majarian *et al.*, 1989), was made possible by the cloning and sequencing of the structural genes for several *Salmonella* flagellins corresponding to non-crossreactive flagellar antigens (Joys, 1985; Wei and Joys, 1985). Comparison of the sequences of the ca. 1500 base pair sequences of the different alleles showed them to be identical or nearly identical for several hundred base pairs at each end of the genes, but widely diverse in the middle, with region IV, of ca. 360 base pairs, showing no more than 30% amino acid sequence homology in any pairwise comparison. The presence of two EcoRV sites, 48 base pairs apart, in region IV of the flagellin gene of *S. muenchen* (H antigen d) allowed the *in vitro* deletion of a 48 base-pair segment and the insertion by blunt-end ligation of a 45 base-pair synthetic oligonucleotide specifying a peptide epitope, CTP3, of the B subunit of *V. cholerae* toxin (Jacob *et al.*, 1983). An aromatic-dependent strain of *S. dublin* made flagellin-negative by inactivation of its only flagellin gene became motile when given the chimeric flagellin gene, which indicates that the insertion did not interfere with correct assembly of flagellin into flagellar filaments. Exposure of the foreign epitope at the surface of the flagellar filaments was shown by the effect of a monoclonal antibody against CTP3, that is, bacterial immobilization and immunogold electron microscope labelling. Mice given three doses, i.p., of the live vaccine strain with the recombinant flagellin gene made antibody reactive by ELISA with either CTP3 peptide or whole cholera toxin as coating antigen (Newton *et al.*, 1989). The same procedure was used to obtain expression of two different epitopes of hepatitis B surface antigen in flagellin (Wu *et al.*, 1989). Only one of four plasmids with at least one copy of the relevant oligonucleotide inserted in the desired orientation conferred motility. However, the recombinant flagellins specified by the other three chimeric genes were detected, both by Western blot of cell lysates and by the immune response of mice, guinea pigs and rabbits given the live vaccine strain carrying any one of the four plasmids concerned, as tested by ELISA, with the relevant synthetic peptide or whole protein as coating antigen. Expression of several other epitopes has been obtained by the same procedure (Newton *et al.*, 1990). In experiments in collaboration with the late Dr. Ed Beachey and his coworkers at the University of Tennessee, Memphis, the serum antibody response (ELISA titer with peptide or M protein as coating antigen, and opsonizing activity) of mice to several injections of a ΔaroA live vaccine strain with flagella made of flagellin with an insert corresponding to the N-terminal 15 amino acids of the M protein of *Streptococcus pyogenes* type 5 were such as to predict protection. Mice given this live vaccine, several doses, by the i.p. route, were found partially protected against a later i.p. *S. pyogenes* challenge which killed mice similarly immunized but with the live vaccine strain with no insert in its flagellin (Newton *et al.*, 1991). This protection was specific in that immunized mice did not survive challenge with a type 24 strain.

The sequences so far tested in this system were chosen in part because they were known to be epitopes, and, in some cases, able to evoke biologically relevant immune responses when presented in other ways, such as in the source protein or as conjugates. The extent to which location in flagellin, and thus proximity to at least B-cell epitopes will confer or improve immunogenicity of unselected peptide sequences remains to be discovered.

References

Akiyama *et al.*, 1964, The use of diffusion chambers in investigating the cellular nature of immunity in experimental typhoid and tuberculosis, *Japan. J. Microbiol.* 8:37-48.

Bacon, G.A., Burrows, T.W. and Yates, M., 1951, The effects of biochemical mutation on the virulence of *Bacterium typhosum*: The loss of virulence of certain mutants, *Br. J. Exp. Pathol.* 32:85-96.

Brown, A. *et al.*, 1987, An attenuated aroA *Salmonella typhimurium* vaccine elicits humoral and cellular immunity to cloned beta-galactosidase in mice, *J. Infect. Dis.* 155:86-92.

Charbit, A., Boulain, J.C., Ryter, A., and Hofnung, M., 1986, Probing the topology of a bacterial membrane protein by genetic insertion of a foreign epitope; expression at the cell surface, *EMBO J.* 5:3029-3037.

Charbit, A. *et al.*, 1988, Expression of a poliovirus neutralization epitope at the surface of recombinant bacteria: First immunization results, *Ann. Inst. Pasteur Microbiol.* 139:45-58.

Clements, T.D. *et al.*, 1986, Oral immunization of mice with attenuated *Salmonella enteritidis* containing a recombinant plasmid which codes for production of the B subunit of heat-labile *E. coli* enterotoxin, *Infect. Immun.* 53:685-692.

Cryz, S.J. *et al.*, 1989, Construction and characterization of a Vi-positive variant of the *Salmonella typhi* live oral vaccine strain Ty21a, *Infect. Immun.* 57:3863-3868.

Curtiss, R., III. and Kelly, S.M., 1987, *Salmonella typhimurium* deletion mutants lacking adenylate cyclase and cyclic AMP receptor protein are avirulent and immunogenic, *Infect. Immun.* 55:3035-3043.

Edwards, M.F. and Stocker, B.A.D., 1988, Construction of ΔaroA hisΔpur strains of *Salmonella typhi*, *J. Bacteriol.* 170:3991-3995.

Fahey, K.J. and Cooper, G.N., 1970, Oral immunization in experimental Salmonellosis II. Characteristics of the immune response to temperature-sensitive mutants given by oral and parenteral routes, *Infect. Immun.* 2:183-191.

Fairweather, N.E. *et al.*, 1990, Oral vaccination of mice against tetanus by use of a live attenuated *Salmonella* carrier, *Infect. Immun.* 58:1323-1326.

Germanier, R., 1970, Immunity in experimental Salmonellosis I. Protection induced by rough mutants of *Salmonella typhimurium*, *Infect. Immun.* 2:309-315.

Germanier, R. and Furer, E., 1971, Immunity in experimental Salmonellosis: II. Basis for the avirulence and protective capacity of galE mutants of *Salmonella typhimurium*, *Infect. Immun.* 4:663-673.

Germanier, R. and Furer, E., 1975, Isolation and chararcterization of galE mutant Ty21a of *Salmonella typhi*: A candidate strain for a live, oral typhoid vaccine, *J. Infect. Dis.* 131:553-558.

Gilman, R.*H. et al.*, 1977, Evaluation of a UDP-glucose-4-epimeraseless mutant of *Salmonella typhi* as a live oral vaccine, *J. Infect. Dis.* 136:717-723.

Herrington, D.A. *et al.*, 1990, Studies in volunteers to evaluate *Shigella* vaccines: Further experience with a bivalent *Salmonella typhi-Shigella sonnei* vaccine and protection conferred by previous *Shigella sonnei* disease, *Vaccine* 8:353-357.

Hoiseth, S.K. and Stocker, B.A.D., 1981, Aromatic-dependent *Salmonella typhimurium* are non-virulent and effective as live vaccines, *Nature* 291:238-239.

Hoiseth, S.K. and Stocker, B.A.D., 1985, Genes aroA and serC of *Salmonella typhimurium* constitute an operon, *J. Bacteriol.* 163:355-61.

Hone, D.M. *et al.*, 1988, A galE via (Vi antigen-negative) mutant of *Salmonella typhi* strain Ty2 retains virulence in humans, *Infect. Immun.* 56:1326-1333.

Jacob, C.O., Sela, M. and Arnon, R., 1983, Antibodies against synthetic peptides of the B subunit of cholera toxin: Cross reaction and neutralization of the toxin. *Proc. Natl. Acad. Sci. U.S.A.* 80:7611-7615.

Joys, T.M., 1985, The covalent structure of the phase-1 flagellar filament protein of *Salmonella typhimurium* and its comparison with other flagellins. *J. Biol. Chem.* 260:15758-15761.

Levine, M.*M. et al.*, 1976, Attenuated Streptomycin-dependent *Salmonella typhi* oral vaccine: Potential deleterious effects of lyophilization, *J. Infect. Dis.* 133:424-429.

Levine, M.M. *et al.*, 1987, Safety, infectivity, immunogenicity and *in vivo* stability of tuso attenuated auxotrophic mutant strains of *Salmonella typhi*, 541Ty and 543Ty, as live oral vaccines in man, *J. Clin. Invest.* 79:888-902.

Levine, M.D. *et al.*, 1990, New and improved vaccines against typhoid fever, 269-287, In "New Generation Vaccines," edit. Woodrow, C. and Levine, M.M., Academic Press, London.

Linde, K., 1983, Stable high immunogenic mutants of *Salmonella* with two independent attenuating markers as potential live vaccine, *Devel. Biol. Stand.* 53:15-28.

Linde, K., Beev, J., and Bondarenki, V., 1990, Stable *Salmonella* live vaccine strains with two or more attenuating mutations and any desired level of attenuation, *Vaccine* 8:278-282.

Majarian, W.R., Kasper, S.J. and Brey, R.N., 1989, Expression of heterologous epitopes as recombinant flagella on the surface of attenuated *Salmonella*, in "Vaccines 89: Modern approaches to new vaccines including prevention of AIDS" (ed. R.A. Lerner et al.), (p. 277-281) Cold Spring Harbor Laboratory, New York.

Maskell, D.J. *et al.*, 1987, *Salmonella typhimurium* aroA mutants as carriers of the *E. coli* heat-labile enterotoxin B subunit to the murine secretory and systemic immune systems, *Microb. Pathog.* 2:211-221.

McFarland, W.C. and Stocker, B.A.D., 1987, Effect of different purine auxotrophic mutations on mouse-virulence of a Vi-positive strain of *Salmonella dublin* and of two strains of *Salmonella typhimurium*, *Microb. Pathog.* 3:129-141.

Morris, J.A., Wray, C. and Sojka, W.J., 1976, The effect of T and B lymphocyte depletion on the protection of mice vaccinated with a galE mutant of *Salmonella typhimurium*, *Brit. J. Exper. Path.* 57:354-360.

Mukkur, T.K. *et al.*, 1987, Protection against experimental salmonellosis in mice and sheep by immunization with aromatic-dependent *Salmonella typhimurium*, *J. Med. Microbiol.* 24:11-19.

Newton, S.M.C., Jacob, C.O. and Stocker, B.A.D., 1989, Immune response to cholera toxin epitope inserted in *Salmonella* flagellin, *Science* 244:70-72.

Newton, S.M.C., Manning, W.C., Hovi, M., and Stocker, B.A.D., 1990, Aromatic-dependent *Salmonella* with foreign epitope insert in flagellin as live vaccine. In "Vaccines 90: Modern Approaches to New Vaccines Including Prevention of AIDS", (ed. F. Brown *et al.*), (pp. 439-445) Cold Spring Harbor Laboratory, New York.

Newton, S.M.C. *et al.*, 1991, Expression and immunogenicity of a Streptococcal M protein epitope inserted in *Salmonella* flagellin, *Infect. Immun.* 59: 2158-2165.

Nnalue, N.A. and Stocker, B.A.D., 1986, Some galE mutants of *Salmonella choleraesuis* retain virulence, *Infect. Immun.* 54:635-640.

O'Callaghan, D. *et al.*, 1988, Characterization of aromatic- and purine-dependent *Salmonella typhimurium*: Attenuation, persistence and ability to induce protective immunity in BALB/c mice, *Infect. Immun.* 56:419-423.

Poirier, T.P., Kehoe, M.A., and Beachey, E.H., 1988, Protective immunity evoked by oral administration of attenuated aroA *Salmonella typhimurium* expressing cloned streptococcal M protein, *J. Exp. Med.* 168:25-32.

Robbins, J.B. and Schneerson, R., 1990, Polysaccharide-protein conjugates: A new generation of vaccines, *J. Infect. Dis.* 161:821-832.

Robertsson, J.A. *et al.*, 1983, *Salmonella typhimurium* infection in calves: Protection and survival of virulent challenge bacteria after immunization with live or inactivated vaccines, *Infect. Immun.* 41:742-750.

Sigwart, D.F., Stocker, B.A.D. and Clements, J.D., 1989, Effect of a purA mutation on the efficacy of *Salmonella* live-vaccine vectors. *Infect. Immun.* 57:1858-1861.

Smith, B.P. *et al.*, 1984, Aromatic-dependent *Salmonella typhimurium* as modified live vaccines for calves, *Am. J. Vet. Res.* 45:59-66.

Smith, H.W., 1965, The immunization of mice, calves and pigs against *Salmonella dublin* and *Salmonella chloraesuis* infections. *J. Hyg. (Camb.)* 63:117-135.

Stocker, B.A.D., 1988, Auxotrophic *Salmonella typhi* as live vaccine, *Vaccine* 6:141-145.

Tacket, C.O. *et al.*, 1990, Safety, immunogenicity, and efficacy against cholera challenge of a typhoid-cholera hybrid derived from *Salmonella typhi* Ty21a, *Infect. Immun.* 58:1620-1627.

Tite *et al.*, 1988, Antiviral immunity induced by recombinant nucleoprotein of influenza A virus. I. Characterization and cross-reactivity of T cell responses, *J. Immunol.* 141:3980-3987.

Wahdan, M.H. *et al.*, 1982, A controlled field trial of livem *Salmonella typhi* strain Ty21a oral vaccine against typhoid: Three year results, *J. Infect. Dis.* 145:292-295.

Wei, L.-N. and Joys, T.M., 1985, Covalent structure of three phase-1 flagellar filament proteins of *Salmonella*, *J. Mol. Biol.* 186:791-803.

Wu, J.Y. *et al.*, 1989, Expression of immunogenic epitopes of hepatitis B surface antigen with hybrid flagellin proteins by vaccine strain of *Salmonella*, *Proc. Natl. Acad. Sci. U.S.A.* 86: 4726-4730.

Molecular and Cellular Basis of Eucaryotic Cell Invasion by *Shigella flexneri*

Philippe J. Sansonetti

Introduction

Bacteria belonging to the genus *Shigella* cause bacillary dysentery, an invasive disease of the human colon. This disease is prevalent in tropical regions, particularly in some overcrowded areas of the developing world. The essential characteristics of the pathogenic potential of *Shigella* reside in its capacity to invade cells of the colonic epithelium (LaBrec *et al.*, 1964; Takeuchi *et al.*, 1968). This communication will review recent data on the molecular and cellular basis of this invasive process. "Invasion" is a general term that summarizes several stages of interaction of the bacterium with its host cell such as entry into epithelial cells, intracellular multiplication, intracellular movement and cell to cell spread, and eventually death of the host-cell (Maurelli *et al.*, 1988). This sequence of events leads to an amplification of the bacterial inoculum within the intestinal epithelium which allows subsequent passage of numerous bacteria into the lamina propria (the connective tissue of the intestinal villus) and causes a strong inflammatory reaction which accounts for multiple abscesses ulcerated within the intestinal lumen, thus for the dysenteric symptoms characteristic of the disease: fever, abdominal cramps and tenesmus, bloody and mucopurulent stools.

Bacterial invasion of epithelial cells which are not professional phagocytes like polymorphonuclears or monocytes/macrophages, can be studied by infecting continuous mammalian cell lines such HeLa or Henle cells (LaBrec *et al.*, 1964; Hale *et al.*, 1979a; Hale *et al.*, 1979b). Tissue invasion can be studied in more definitive assays such as the Sereny test in which invasive bacteria elicit a strong keratoconjunctivitis in guinea pigs (Sereny, 1957) or in the rabbit ligated ileal loop (Formal *et al.*, 1961). However, the only animal species which reproduces the human disease is macaque monkey following intragastric inoculation. A useful cell assay system called plaque assay (Oaks *et al.*, 1985) has been developed in which infection of a confluent monolayer of epithelial cells allows formation of plaques due to the cytopathic effect of invasive isolates of *S. flexneri*. In this assay, several stages of the cell invasion process can be explored; e.g. entry, intracellular multiplication, cell to cell spread and host cell killing.

Entry of *Shigella flexneri* into Epithelial Cells

Cell Biology of the Entry Process

Transmission electron microscopy studies carried out during infection of cell monolayers have shown that invasive shigellae were internalized via an endocytic process which requires energy production by both the bacteria and the host-cells (Hale *et al.*, 1979a; Hale *et al.*, 1979b). It is inhibited by cytochalasins, a family of molecules that block polymerization of actin subunits (G-actin) in actin filaments (F-actin) (Tanenbaum, 1978). This indicates that a process similar to phagocytosis is involved (Stendahl *et al.*, 1980; Sheterline *et al.*, 1984). Formation of an actino-myosin complex has been directly demonstrated at the site of bacterial entry within HeLa cells by double immunofluorescence labelling using NBD-phallacidin, a fluorescent dye specific for F-actin (Barak, 1980) and a monoclonal antibody directed against myosin. These experiments have demonstrated the accumulation of these two major components of the host-cell cytoskeleton underneath the cytoplasmic membrane at the site of bacterial entry (Clerc *et al.*, 1987). *S. flexneri* therefore induces cells which are not professional phagocytes to express an endocytic process close to classical phagocytosis.

Host cell receptors involved in this process as well as the transmembrane signals inducing cytoskeleton reorganization are as yet unknown. No significant change in cytosolic calcium concentrations (Ca^{++}) has been observed during synchronized bacterial entry into HeLa cells (Clerc *et al.*, 1989). On the other hand, in the case of *Yersinia pseudotuberculosis*, the target of the invasin (Inv), an outer membrane protein responsible for binding to and entry into the host cell (Isberg *et al.*, 1985) is an integrin (Isberg *et al.*, 1988). Integrins are integral transmembrane receptor proteins which, via their extracellular domain, interacts with ligands such as fibronectin and other matrix proteins, via their cytoplasmic domain with the host-cell cytoskeleton through proteins such as talin and vinculin (Hynes, 1987). A model is taking shape in which bacteria, or more specifically domains of bacterial surface proteins, may mimic domains of matrix proteins, bind to integrins and trigger entry as recently shown for of the FHA adhesin of *Bordetella pertussis* which interacts with the CR3 receptor of macrophages (Relman *et al.*, 1990). It remains to be demonstrated that this model can be generalized to other invasive species, especially *Shigellae*. Perhaps other mechanisms should also be considered. We have recently shown accumulation of clathrin underneath the cytoplasmic membrane at the entry site of the bacterium (Clerc *et al.*, 1989). This 180 kD protein is the major component which surrounds and stabilizes coated pits and coated vesicles during receptor mediated endocytosis (Pearse *et al.*, 1976; Goldstein *et al.*, 1979). This alternative mechanism to phagocytosis allows endocytosis of various molecules such as receptors, hormones, toxins or even viruses. Conditions blocking clathrin polymerization prevent entry of *S. flexneri* (Clerc *et al.*, 1989). These data suggest that this microorganism can mobilize all the major endocytic pathways available in the cell.

Genetic and Molecular Studies of *S. flexneri* Invasins

The genes responsible for the invasive phenotype are carried by a 220 kb plasmid in *S. flexneri* (Sansonetti *et al.*, 1982). Homologous plasmids are present in all invasive isolates from the four *Shigella* species (Sansonetti *et al.*, 1983). These genes have been cloned in a cosmid vector (Maurelli *et al.*, 1985). Tn5 mutagenesis has identified five loci on a 25 kb sequence (Baudry *et al.*, 1987). This is in agreement with studies carried out by other groups in different serotypes and species (Watanabe *et al.*, 1986; Sasakawa *et al.*, 1988). One of these five loci has been extensively studied. "Locus 2" contains genes encoding four polypeptides which are known to be particularly immunogenic (Hale *et al.*, 1985), based on their consistent recognition by sera from convalescing patients or monkeys. IpaA (78 kD), IpaB (62 kD), IpaC (43 kD) and IpaD (38 kD) are encoded by contiguous genes which are organized as an operon in addition with genes encoding a 21 and 17 kD polypeptide in the following order from 5'to 3': 21 kD (Ipp21), 17 kD (Ipp17), IpaB, C, D, and A, (Baudry *et al.*, 1987; Baudry *et al.*, 1988; Buysse *et al.*, 1987; Sasakawa *et al.*, 1989). A weak accessory promoter exists before IpaD (Figure 1). No signal sequence has been detected at the NH_2 terminus of any of these six polypeptides. We have recently studied the function of each of the four Ipas by gene destruction in pWR100, the 220 kb virulence plasmid of *S. flexneri* serotype 5 (M90T) (High, Sansonetti, submitted). Each ipa gene has been cloned individually and deleted from a central sequence. A cassette encoding resistance to spectinomycin has been subsequently cloned to fill the deletion. These *in vitro* mutated genes have each been subcloned into a suicide vector and exchanged with the wild type gene by double-allelic recombinations. In order to avoid the polar effect of these mutations, a promoter has been positioned downstream the mutated genes to allow further transcription of genes in the operon. Phenotypes of the four mutations are as follows: *ipa*B and *ipa*C mutants are no longer invasive but express a strong ability to bind to HeLa cells. IpaB and C are therefore considered as the "invasins" of *S. flexneri*. These mutations also allow us to separate an adhesion and an entry step. The *ipa*D mutant is neither invasive nor adhesive. IpaD may therefore be either the *S. flexneri* adhesin or a protein involved in the organization of the adhesin/entry complex. Finally, *ipa*A mutants are slightly less invasive than the wild type strain. They show a phenotype of intracellular localization within two hours after entry into HeLa cells. However, after this early period, bacteria spread intracellularly and from cell to cell as will be described later (Ics phenotype). As shown in Fig. 1, downstream the *ipa* locus ("locus 2"), "locus 1" has been located

"Locus 1" contains *vir*B (also called *inv*R), a positive regulator of the entry genes (Sasakawa *et al.*, 1989). Functions of proteins encoded by genes in loci

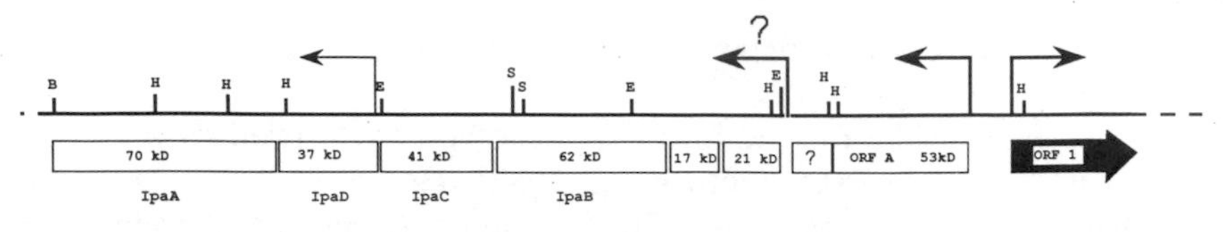

Figure 1. Map of Virulence Plasmid "Locus 2"

3, 4 and 5 are not yet characterized. One of them, *mxi* has been proposed as necessary for proper translocation and positioning of Ipa proteins during the entry process (Hromockyj *et al.*, 1989). Existence of such genes appears necessary, due to the lack of signal peptides identified at the level of Ipas.

In summary, the genetic and molecular analysis of *S. flexneri* entry into cells is still at its early stage. Current research is also active in the field of environmental regulation of entry.

Intracellular Multiplication of *Shigella*

A major characteristic of the intracellular behavior of *S. flexneri* is its tremendous capacity to multiply intracellularly. Intracellular growth starts within minutes after entry and brings the intracellular population from 5 to an average of 500 bacteria per infected cell in 5 hours (Sansonetti *et al.*, 1986). Over this period, *S. flexneri* appears quite different from *Salmonella* and *Yersinia*. These other enteroinvasive species grow slowly in the intracellular medium (Sansonetti *et al.*, 1986; Small *et al.*, 1987; Finlay *et al.*, 1988). In *Salmonella*, a long lag period (6-8h) is observed before intracellular growth starts (Finlay *et al.*, 1988). *Yersinia* survives intracellularly but its growth capacities are very limited. Rapid and efficient intracellular growth of *Shigella* is certainly a critical aspect of its pathogenicity as compared with other enteroinvasive microorganisms. One can speculate that *Shigella* has elected the intestinal epithelial cell as its site of multiplication, thus making shigellosis essentially an epithelial disease. Neither production of Shiga-toxin or of a Shiga-like toxin (Clerc *et al.*, 1987; Fontaine *et al.*, 1988) nor the production of aerobactin (Nassif *et al.*, 1987; Lawlor *et al.*, 1987) account for rapid intracellular growth. On the other hand, sequential transmission electron microscopy studies have demonstrated that *S. flexneri* lysed its phagocytic vacuole early after entry (Sansonetti *et al.*, 1986) unlike *Salmonellae* and *Yersiniae*. The latter must adapt to the hostile environment of the phagocytic vacuole, particularly once phagolysosomal fusion has occurred.

Intracellular Movement and Cell-to-Cell Spread of *S. flexneri*

Molecular and Cellular Basis

Although *S. flexneri* does not express flagella, once intracellular, its shows rapid and random movements which allow occupation of the entire cytoplasmic volume. These movements also allow cell-to-cell spread on confluent monolayers and, most probably, in epithelia. This process permits epithelial colonization without contact with the extracellular medium. Characteristics of this movement have been described more than two decades ago by time lapse microcinematographic observation (Ogawa *et al.*, 1968). More recently, a mutant has been obtained which was described as unable to pass from one cell to another while invading the guinea pig's cornea in the Sereny test (Makino *et al.*, 1986). The corresponding gene has been called *vir*G. It is located on the virulence plasmid of *S. flexneri* and codes for a 120 kD outer membrane protein (Lett *et al.*, 1989). Molecular basis of this movement have recently been analyzed in more detail (Pal *et al.*, 1989; Bernardini *et al.*, 1989). Staining of infected cells with the fluorescent dye NBD-phallacidin has shown that intracellular bacteria were often covered with F-actin and that some of these bacteria were even followed by a thick tail of polymerized actin (Bernardini *et al.*, 1989). Some of these tails, with a bacterium at their tip, formed protrusions which ensured passage of the bacteria from one cell to another without "seeing" the extracellular medium. It has recently been shown that *Listeria monocytogenes* displays a similar behavior while infecting macrophages or the enterocyte-like cell line Caco-2 (Tilney *et al.*, 1989; Mounier *et al.*, 1990). The following sequence is therefore observed: bacterium in intracellular position - lysis of the phagocytic vacuole (contact hemolytic activity in *S. flexneri*, listeriolysin 0 in *L. monocytogenes*) - interaction with the host-cell cytoskeleton - elicitation of multiple foci of actin nucleation and polymerization at the bacterial surface followed by polarization of the gel - generation of the movement - formation of a protrusion into an adjacent cell - presence of the bacterium bound within a double membrane - lysis of the double membrane - presence of the bacterium free within the next cell - and so on... In *S. flexneri*, a transposon mutant has been obtained which is unable to make plaques on confluent HeLa cells (Bernardini *et al.*, 1989). This mutant forms microcolonies which cap the nucleus in infected cells. It does not move intracellularly, is unable to spread from one cell to another although it still lyses its phagocytic vacuole, and is subsequently negative in the Sereny test. The identified plasmid gene (*Ics*A) is equivalent to the previously described *vir*G (Makino *et al.*, 1986). The function of its 120 kD product which interacts with the host cell cytoskeleton is currently under study.

In vivo Significance of the Ics Phenotype of *S. flexneri*

After intragastric infection of macaque monkeys with a *ics*A mutant of *S. flexneri*, animals do not develop significant symptoms of dysentery as compared with animals infected with the wild type strain. Endoscopic examination of the rectum and sigmoid colon shows only slightly ulcerated lesions dispersed on the mucosa whereas in the case of animals infected with the wild type strain, many coalescing abscesses, hemorragic ulcerations and purulent membranes can be observed (Sansonetti *et al*, submitted). When the abscesses caused by the *ics*A mutant are biopsied, they appear essentially located over lymphoid follicles. These observations confirm the major role that the Ics phenotype plays in the pathogenesis of shigellosis. It is likely that invasion of the colonic epithelium starts at sites located over lymphoid follicles which may be equivalent to Peyer's patches. These sites are expected to be rich in M cells (Bye *et al.*, 1984). Utilization of M cells by invasive microorganisms must be the price paid to maintain efficient sampling of luminal antigens in order to elicit efficient mucosal immune responses. A suggested process of epithelial invasion by *S. flexneri* is summarized in Figure 2.

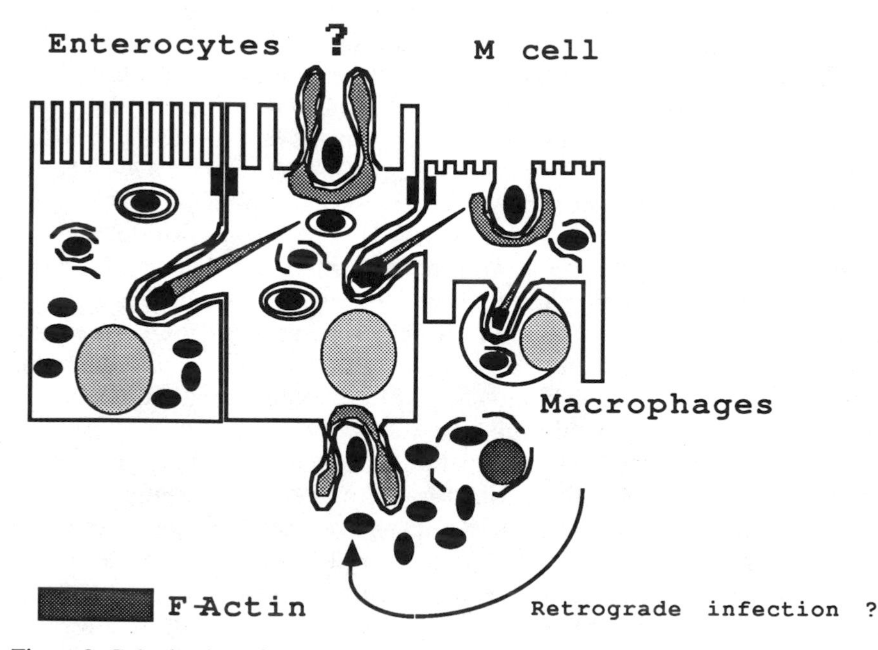

Figure 2. Colonization of the Intestinal Epithelium by *Shigella spp.*

Shigella is a good model to study invasion of eucaryotic cells and tissues. This model can also bring informations at a more fundamental level on topics such as phagocytosis or regulation of the cytoskeleton's structure. On the other hand, practical applications are also expected, such as the design of live vaccine strains administered orally to elicit protective mucosal response.

References

Barak, L.S., *et al.*, 1980, Fluorescence staining of the actin cytoskeleton in living cells with 7-nitrobenz-2-oxa-1,3-diazole-phallacidin, *Proc. Natl. Acad. Sci. USA*. 77: 980-984.

Baudry, B., Kaczorek, M., and Sansonetti, P.J., 1988, Nucleotide sequence of the invasion plasmid antigen B and C genes (IpaB and IpaC) of *Shigella flexneri, Microb. Pathogen.* 4: 345-357.

Baudry, B., *et al.*, 1987, Localization of plasmid loci necessary for the entry of *Shigella flexneri* into HeLa cells, and characterization of one locus encoding four immunogenic polypeptides, *J. Gen. Microbiol.* 133: 3409-3413.

Bernardini, M.L., *et al.*, 1989, Identification of IcsA, a plasmid locus of *Shigella flexneri* that governs intra- and intercellular spread through interaction with F-actin, *Proc. Natl. Acad. Sci. USA*, 86: 3867-3871.

Buysse, J.M., *et al.*, 1987, Molecular cloning of invasion plasmid antigen (Ipa) genes from *Shigella flexneri*: Analysis of Ipa gene products and genetic mapping, *J. Bacteriol.* 169: 2561-2569.

Bye, W.A., Allan, C.H., and Trier, J.S., 1984, Structure, distribution, and origin of M cells in Peyer's patches of mouse ileum, *Gastroenterology*, 86: 789-801.

Clerc, P., and Sansonetti, P.J., 1987, Entry of *Shigella flexneri* into HeLa cells: Evidence for directed phagocytosis involving actin polymerization and myosin accumulation, *Infect. Immun.* 55: 2681-2688.

Clerc, P., and Sansonetti, P.J., 1989, Evidence for clathrin mobilization during directed phagocytosis of *Shigella flexneri* by Hep2 cells, *Microbial. Pathogenesis*, 7: 329-336.

Clerc, P., *et al.*, 1987, Plasmid mediated early killing of eucaryotic cells by *Shigella flexneri* as studied by infection of J774 macrophages, *Infect. Immun.* 55: 521-527.

Clerc, P., *et al.*, 1989, Internalization of *Shigella flexneri* into HeLa cells occurs without an increase in Ca^{++} concentration, *Infect. Immun.* 57: 2919-2922.

Finlay, B.B., and Falkow, S., 1988, A comparison of microbial invasion strategies of *Salmonella, Shigella* and *Yersinia* species, *UCLA Symp. Mol. Cell. Biol.* 64: 227-243.

Finlay, B.B., Gumbiner, B., and Falkow, S., 1988, Penetration of *Salmonella* through a polarized Madin-Darby canine kidney epithelial cell monolayer, *J. Cell. Biol.* 107: 221-230.

Fontaine, A., Arondel, J., and Sansonetti, P.J., 1988, Role of Shiga toxin in the pathogenesis of shigellosis as studied using a Tox-mutant of *Shigella dysenteriae* 1, *Infect. Immun.* 56: 3099-3109.

Formal, S.B., Kundel, D., Schneider, H., Kunev, N., and Sprinz, H., 1961, Studies with *Vibrio cholerae* in the ligated loop of the rabbit intestine, Br. *J. Exp. Pathol.* 42: 504-510.

Goldstein, J.L., Anderson, R.G.W., and Brown, M., Coated pits, coated vesicles, and receptor-mediated endocytosis, *Nature*. 279: 679-685.

Hale, T.L., and Bonventre, P.F., 1979(a), *Shigella* infection of Henle intestinal epithelial cells: role of the bacteria, *Infect. Immun.* 24: 879-886.

Hale, T.L., Morris, R.E., and Bonventre, P.F., 1979(b), *Shigella* infection of Henle intestinal epithelial cells: role of the host cell, *Infect. Immun.* 24: 887-894.

Hale, T.L., Oaks, E.V., and Formal, S.B., 1985, Identification and antigenic characterization of virulence-associated, plasmid-coded proteins of *Shigella spp* and enteroinvasive *Escherichia coli*, *Infect. Immun.* 50: 620-629.

Hromockyj, A.E., and Maurelli, A.T., 1989, Identification of *Shigella* invasion genes by isolation of temperature-regulated inv::lacZ operon fusions, *Infect. Immun.* 57: 2963-2970.

Hynes, R.D., 1987, Integrins, a family of cell surface receptors, *Cell*, 48: 549-554.

Isberg, R.R., and Falkow, S., 1985, A single genetic locus encoded by *Yersinia pseudotuberculosis* permits invasion of cultured mammalian cells by *E. coli* K12, *Nature*, 317: 262-264.

Isberg, R.R., and Leong, J.M., 1988, Cultured mammalian cells attach to the invasin protein of *Yersinia pseudotuberculosis*, *Proc. Natl. Acad. Sci. USA*, 85: 6682-6686.

LaBrec, E.H., *et al.*, 1964, Epithelial cell penetration as an essential step in the pathogenesis of bacillary dysentery, *J. Bacteriol.* 88: 1503-1518.

Lawlor, K.M., *et al.*, 1987, Virulence of iron transport mutants of *Shigella flexneri* and utilization of host iron compounds, *Infect. Immun.* 55: 594-599.

Lett, M., *et al.*, 1989, VirG, a plasmid-coded virulence gene of *Shigella flexneri*: Identification of the VirG protein and determination of the complete coding sequence, *J. Bacteriol.* 171: 353-359.

Makino, S., *et al.*, 1986, A genetic determinant required for continuous reinfection of adjacent cells on large plasmid in *Shigella flexneri* 2a, *Cell.* 46: 551-555.

Maurelli, A.T., and Sansonetti, P.J., 1988, Genetic determinants of *Shigella* pathogenicity, *Ann. Rev. Microbiol.* 42: 127-150.

Maurelli, A.T., *et al.*, 1985, Cloning of plasmid DNA sequences involved in invasion of HeLa cells by *Shigella flexneri*, *Infect. Immun.* 49: 164-171.

Mounier, J., *et al.*, 1990, Intracellular and cell-to-cell spread of *Listeria monocytogenes* involves interaction with F-actin in the enterocytelike cell line Caco 2, *Infect. Immun.* 58: 1948-1058.

Nassif, X., *et al.*, 1987, Evaluation with an iuc::Tn10 mutant of the role of aerobactin production in the virulence of *Shigella flexneri*, *Infect. Immun.* 48:124-129.

Oaks, E.V., Wingfield, M.E., and Formal, S.B., 1985, Plaque formation by virulent *Shigella flexneri*, *Infect. Immun.* 53: 57-63.

Ogawa, H., Nakamura, A., and Nakaya, R., 1968, Cinematographic study of tissue cell cultures infected with *Shigella flexneri*, *Japan J. Med. Sci. Biol.* 21: 259-273.

Pal, T., *et al.*, 1989, Intracellular spread of *Shigella flexneri* associated with the kcpA locus and a 140 kilodalton protein, *Infect. Immun.* 57: 477-486.

Pearse, B.M.F., 1976, Clathrin: A unique protein associated with intracellular transfer of membrane by coasted vesicles, *Proc. Natl. Acad. Sci. USA*, 73: 1255-1259.

Sansonetti, P.J., Kopecko, D.J., and Formal, S.B., 1982, Involvement of a plasmid in the invasive ability of *Shigella flexneri*, *Infect. Immun.* 35: 852-860.

Sansonetti, P.J. *et al.*, 1983, Molecular comparison of virulence plasmids in *Shigella* and enteroinvasive, *Ann. Microbiol. (Paris)*. 134A: 295-318.

Sansonetti, P.J., *et al.*, 1986, Multiplication of *Shigella flexneri* within HeLa cells: Lysis of the phagocytic vacuole and plasmid mediated contact hemolysis, *Infect. Immun.* 55: 521-527.

Sasakawa, C., *et al.*, 1988, Virulence-associated genetic region comprising 31 kilobases of the 230 kilobase plasmid in *Shigella flexneri* 2a, *J. Bacteriol.* 170: 2480-2484.

Sasakawa, C., *et al.*, 1989, Functional organization and nucleotide sequence of virulence region-2 on the large virulence plasmid in *Shigella flexneri* 2a, *Mol. Microbiol.* 3: 1191-1201.

Sereny, B., 1957, Experimental keratoconjunctivitis shigellosa, *Acta Microbiol. Acad. Sci. Hung.* 4: 367-376.

Sheterline, P., Richard, J.E., and Richards, J.C., 1984, Fc receptor-directed phagocytic stimuli induce transient actin assembly at an early stage of phagocytosis in neutrophil leukocytes, *Eur. J. Cell. Biol.* 34:80-87.

Small, P.L.C., Isberg, R.R., and Falkow, S., 1987, Comparison of the ability of enteroinvasive *Escherichia coli*, *Salmonella typhimurium*, *Yersinia pseudotuberculosis* and *Yersinia enterocolitica* to enter and replicate within HEp2 cells, *Infect. Immun.* 55: 1674-1679.

Stendahl, O.I., *et al.*, 1980, Distribution of acting-binding protein and myosin in macrophages during spreading and phagocytosis, *J. Cell. Biol.* 84: 215-224.

Takeuchi, A., Formal, S.B., and Sprinz, H., 1968, Experimental acute colitis in the rhesus monkey following peroral infection with *Shigella flexneri.*, *Am. J. Pathol.* 52: 503-519.

Tanenbaum, S.W., 1978, (ed.). Cytocholasins: biochemical and cell biological aspects, North-Holland Publishing Co., Amsterdam.

Tilney, L.G., and Portnoy, D.A., 1989, Actin filaments and the growth movement and spread of the intracellular bacterial parasite, *Listeria monocytogenes*, *J. Cell. Biol.* 109: 1597-1608.

Watanabe, H., and Nakamura, A., 1986. Identification of *Shigella sonnei* form I plasmid genes necessary for cell invasion and their conservation among *Shigella* species and enteroinvasive *Escherichia coli*, *Infect. Immun.* 53: 352-358.

Index

VANDERBILT UNIVERSITY
3 0081 021 626 159